MW00582408

Landing the Paris Climate Agreement

Landing the Paris Climate Agreement

How It Happened, Why It Matters, and What Comes Next

Todd Stern

The MIT Press
Cambridge, Massachusetts
London, England

The MIT Press would like to thank the anonymous peer reviewers who provided comments on drafts of this book. The generous work of academic experts is essential for establishing the authority and quality of our publications. We acknowledge with gratitude the contributions of these otherwise uncredited readers.

This book was set in ITC Stone Serif Std and ITC Stone Sans Std by New Best-set Typesetters Ltd. Printed and bound in the United States of America.

Library of Congress Cataloging-in-Publication Data

Names: Stern, Todd, author.
Title: Landing the Paris Climate Agreement : how it happened, why it matters, and what comes next / Todd Stern.
Description: Cambridge, Massachusetts : The MIT Press, 2024. | Includes bibliographical references and index.
Identifiers: LCCN 2023057517 (print) | LCCN 2023057518 (ebook) | ISBN 9780262049146 (hardcover) | ISBN 9780262379601 (epub) | ISBN 9780262379618 (pdf)
Subjects: LCSH: United Nations Framework Convention on Climate Change (1992 May 9). Protocols, etc. (2015 December 12) | United Nations Framework Convention on Climate Change (1992 May 9). Protocols, etc. (1997 December 11) | Climatic changes—Law and legislation—International cooperation. | Enforcement measures (International law)—Environmental aspects. | Greenhouse gas mitigation—Law and legislation. | Climatic changes—Government policy. | Conference of the Parties (United Nations Framework Convention on Climate Change) (21st : 2015 : Paris, France)
Classification: LCC K3585.5.A42015 S74 2024 (print) | LCC K3585.5.A42015 (ebook) | DDC 344.04/6342—dc23/eng/20240409
LC record available at https://lccn.loc.gov/2023057517
LC ebook record available at https://lccn.loc.gov/2023057518

10 9 8 7 6 5 4 3 2 1

To Jen, Jacob, Zachary, and Ben,
my inspiration, my support, my joy, always

*

To my team at the State Department, the best anywhere,
who made it happen and made it fun

*

To my colleagues in countries around the world who helped, over seven years, bring the Paris Agreement in for a landing, and the memory of two climate heroes and friends who were instrumental in making the Paris Agreement possible—Pete Betts and Tony de Brum

Contents

Prologue

It suddenly hit me shortly after noon on December 12, 2015, the last day of the Paris climate conference, as I sat in the Seine plenary hall next to Secretary of State John Kerry and listened to French foreign minister Laurent Fabius calmly lay out the plan for a day meant to culminate in the adoption of the Paris Agreement. It hit me, somehow for the first time, that we were actually, against long odds, going to succeed in bringing 195 nations into a landmark agreement to contain the metastasizing threat of climate change. Finally, depending on how you count it—after fourteen days of frenetic negotiating, arguing, and cajoling in Paris; or after four years of strategic and tactical maneuvering since the mandate for the Paris negotiation was hammered out in Durban, South Africa; or after seven years of full-throttle effort since I took the reins of a dedicated and talented US negotiating team for a new president; or after twenty years of trying by the countries of the world to negotiate a concrete, effective, operational climate agreement. That's when it hit me. And that's when I got the lump in my throat.

My tenure as special envoy for climate change began on January 26, 2009, our youngest son's fourth birthday, at a ceremony in the ornate Benjamin Franklin room on the eighth floor of the State Department, with that little boy, his two older brothers, and my wife sitting in the front row. The room was full, and there was a palpable sense of optimism in the air about US reengagement on climate change after the eight lean years of the Bush administration. Secretary Hillary Clinton made a strong statement of commitment and resolve, I followed with one of my own, and the message was clear: in this administration, climate change would be a priority and the United States was back. Then without further ado, the preliminaries were over. On my drive home from State, still in the brief afterglow of the event, my cell phone rang, my first official call. It was Al Gore. After graciously congratulating me on the new appointment, the vice president got down to business. He wanted to know, in sum and substance, that I was ready to go all out—I was—and understood the stakes—I did.

* * *

I came to climate change in a roundabout manner. In late 1987, already in my thirties, I swerved away from life as an unenthusiastic lawyer to politics—something I'd wanted to do for a long time—signing on to work in the Michael Dukakis presidential campaign. That's where I met John Podesta in one of those "blink" moments that can change your life. I was already at campaign headquarters in Boston when he showed up one Saturday in early summer 1988 as the incoming head of the new rapid response team. He asked me to have coffee, I told him I had already committed to the vice presidential side of the campaign, he persisted, so I went with him. An hour later, sure that I should stick with this guy, I changed my plans—one of my best decisions ever. After Dukakis lost, my association with John led in relatively short order to working with him and his brother Tony at their political consulting firm, working on Senator Patrick Leahy's (D-VT) Judiciary Committee staff, working for the Clinton transition team, and then joining John in the White House in 1993 as his deputy in the staff secretary's office.[1] In 1995, when John left (he would return and eventually become President Bill Clinton's chief of staff), I moved up to be staff secretary, and that's what I was doing when Clinton's then chief of staff Erskine Bowles found me in 1997 to ask that I borrow some time from my day job to join the team preparing for the December climate conference in Kyoto. At that point, I had nearly ten years of political training under my belt, and with it, a feel for the way politics and policy blend together in Washington, as well as an understanding of the way the White House nerve center relates to the rest of the executive branch. All of that proved invaluable as I got drawn deeper and deeper into climate change, and Erskine's "borrowed time" became the job that never let me go.

With that new assignment in July 1997, my climate change education began. I managed our outreach to stakeholders and climate change communications—which included bringing over a hundred TV weather forecasters to the White House to meet the president and vice president and report back to the folks at home from the North Lawn—as well as joining the senior policy team advising President Clinton on the greenhouse gas emissions target the United States should take into the Kyoto negotiations. I learned about climate science and policy and got schooled in hard-nosed US climate politics, managing the one concern, I used to joke, that brought business and labor together because they both hated my issue. I also got my first exposure to the world of international climate negotiations, participating in the Kyoto conference as the White House adviser on a team led by Stu Eizenstat, then an undersecretary of state. This was my introduction to the semicontrolled chaos of a Conference of the Parties (COP), attended by all 190-plus country parties to the UN Framework Convention on Climate Change (UNFCCC.)[2] I spent most of my time in Kyoto with Stu and his team in a large conference room where representatives from Europe, the United States, and Japan negotiated over their respective emission reduction commitments and related provisions of the protocol. I also got my first taste of the round-the-clock rhythm of the final days of a COP. In the wee hours of the last all-night session, during a break in the US-EU-Japan action, I wandered through the Kyoto International Conference Center past negotiators collapsed on chairs and benches, sprawled on couches, curled up on carpets. It looked like a scene from a Bruegel painting.

In summer 2008, Barack Obama recruited Podesta to run his transition, and John once again brought me in as his deputy—a post that gave me a voice in developing climate policy, among other obligations. As I thought about where I might like to land if Obama won, the idea of leading the US climate negotiating effort at State was at the top of my list. After Obama asked Hillary Clinton to be his secretary of state and she agreed, the stars aligned, since I knew Clinton well from my White House days—both from my work there and because I had the great good fortune one spring day in 1994 to cross paths with her domestic policy adviser, Jennifer Klein, and seventeen months later we were married in the Brooklyn Botanic Garden. Which is to say that I had married into Hillaryland.

A few weeks before the inauguration, after I learned that Clinton wanted me to serve as the special envoy for climate change, I started to think about the team I wanted at State and reached out to Jonathan Pershing and Sue Biniaz, both veteran State Department diplomats whom I knew from my White House days. Jonathan had long since moved on from State and in 2008 was working at the World Resources Institute, an environmental think tank in Washington. I wanted him to be my deputy. When Clinton asked me before she was confirmed whether there was anything I needed for the job, I said, "One thing: Jonathan Pershing." Jonathan is a bearded, bespectacled, long-striding climate polymath—a one-stop shop on all things climate from negotiations to technology to science—who speaks in full paragraphs, is high value-added in most any discussion, and is irrepressibly buoyant despite the gravity of our subject. I knew how much I still did not know about the operational elements of the climate negotiation process, and Jonathan knew that world backward and forward. He would lead the team at the many annual meetings that take place below the so-called ministerial level, and he was known, liked, and respected around the world.

Sue was still at State in the legal adviser's office. I knew her to be a brilliant lawyer, but I didn't know the half of it. She had worked on climate change since the antediluvian days of the Framework Convention, knew what every provision in every agreement meant, is a keen strategist, hardheaded negotiator, lucid writer, and has a Zen-like talent for finding minimalist solutions for intractable snags—with a few words, with commas, even with quotation marks. She often follows a predictable arc from Eeyore-like despondency in the face of impasse to bursts of creative insight, walks slow but thinks fast, and is sardonic, funny, and tough as nails. Outfitted in her daily uniform of a long, loose-fitting denim skirt, Ralph Lauren button-down shirt and clogs, hair in a perpetual bun, she became a tough-love leader to our mostly young climate team, and more important, my own confidant, strategic adviser and partner in crime. Her fingerprints are all over the Paris Agreement and the steps taken in the seven years leading up to it. We would never have gotten there without her.

So with Jonathan lined up, Sue on board, and a team of committed people in the permanent climate office at State eager to serve a new president and new secretary of state, we were ready to go.

Introduction

The 2015 Paris Agreement on climate change was the most important international agreement of the young century and one of the most significant of the past hundred years. It addresses a growing threat that Barack Obama, at a 2015 meeting of business leaders, described as "the only issue other than nuclear weapons that has the capacity to alter the course of human progress." It was a stirring success after twenty years of failed efforts to produce a strong, operational climate agreement. It embodied hope because it makes possible the prospect of transforming the global energy system at something like the speed and scale needed. And it was hard-won given that there are some 195 countries in the UN climate body arrayed in a wide variety of groups, each with their own often conflicting agendas, the substance and politics of climate change are extraordinarily complicated, long-standing North-South resentments aggravate and sometimes paralyze the debate, and negotiations are governed by what amounts to a consensus rule of procedure, requiring that all countries, or nearly all, approve a new agreement.

The success of the Paris negotiation also radiated out beyond climate change. It sent a message about multilateralism itself, about the possibility of solving difficult global problems through common effort, a message of singular importance in a world where most of our greatest challenges are collective action problems. Speaking from the White House on the day the Paris Agreement was adopted, Obama captured this broader message when he said that "together we've shown what's possible when the world stands as one."[1] And this aspect of the Paris Agreement is even more relevant now than it was in 2015, with the impulse toward constructive globalization facing strong headwinds.

The new agreement, reached on December 12, immediately changed the game on climate change, infusing it with a new level of importance, sending a signal around the world to governments, boardrooms, and civil society that climate now commanded the attention of world leaders, who had at last come together to address it. Al Gore, a leading political voice on climate change since the late 1980s, praised "this universal and ambitious agreement," declaring that the transformation to a global clean energy economy "is now firmly and inevitably underway."[2] Michael Levi, the astute climate analyst for the Council on Foreign Relations, noted that "the world finally has a framework for cooperating on climate change that's suited to the task."[3] Of course, there were critics as well, mostly focusing on the nonlegally binding nature of emission commitments, but what the critics wanted was undoable. Diplomacy, as much as domestic politics, is the art of the possible. That understanding is built into the Paris Agreement, but so is the commitment to building a strong and ambitious regime equal to the task.

* * *

The Paris Agreement was built on layers of negotiations that came before, so let's take a step back to set the stage. The original climate treaty was the UN Framework Convention on Climate Change (UNFCCC), concluded in 1992 at the Earth Summit in Rio De Janeiro. The US Senate approved the treaty without controversy on October 7, 1992, toward the end of the first president Bush's administration. The Convention defined its "ultimate objective" as "prevent[ing] dangerous anthropogenic interference with the climate system" and outlined the major steps countries needed to take, including reducing their emissions (mitigation), reporting on their efforts to do that as well as on emission inventories (transparency), building resilience against the impacts of climate change (adaptation), and for developed countries, providing financial and technology support to developing countries. It called on all countries to limit their emissions but did not impose specific obligations, although it articulated a nonbinding aim for developed countries to return their emissions in 2000 to their 1990 levels. Importantly, it also divided the world into two broad categories—Annex 1 for developed countries, and non-Annex 1 for developing, based largely on their material circumstances in 1992. And it set forth a principle of "common but differentiated responsibilities and respective capabilities" (CBDR/RC, or just CBDR in climate jargon), indicating that countries have a common responsibility to act, but that responsibilities and capabilities are not equal.

This principle of differentiation was critical for developing countries, who worried that requirements to reduce their greenhouse gas emissions could disrupt their priorities for growth, development, and the eradication of poverty. Such emissions, after all, are generated by activities crucial to the operation of modern economies—burning fossil fuels to produce electricity, power all forms of transport, heat and cool buildings and homes, and help produce heavy industrial goods, and using land for agriculture and forestry. Moreover, many developing countries saw themselves as getting a raw deal. In their view, developed countries had grown rich on the strength of unfettered access to cheap fossil fuels without worrying about limits on greenhouse gases, so, as a matter of equity, those countries should bear the burden of solving the problem they caused and contribute funds to pay for the costs of developing country efforts to limit emissions and build resilience.

In 1995, the first Conference of the Parties (COP 1) to the Convention was held. Recognizing that the Framework Convention itself could not generate sufficient action to meet its core objective, the parties adopted the "Berlin Mandate" calling for a new agreement by the COP in 1997. That agreement, the Kyoto Protocol, was delivered at COP 3 in Kyoto. For the period 2008–2012, the protocol required that developed countries undertake specific, legally binding obligations to reduce greenhouse gas emissions and follow a set of rigorous rules to track compliance, but, following the explicit dictate of the Berlin Mandate, exempted developing countries from any new commitments. In effect, Kyoto deepened the Framework Convention's division between developed and developing countries, establishing what came to be known as a firewall between them.

But this approach was untenable for two reasons. The first involved the basic math of global emissions. There was no way to meet the Convention's ultimate objective—avoiding dangerous climate change—without significant participation by developing countries since their share of global emissions was already over 50 percent in 1997 and growing rapidly toward 60 percent by 2010.[4] And that share was growing as a percentage of modern emissions, which dwarfed emissions produced long ago—more than half of all global carbon dioxide (CO_2) emissions since 1750 were emitted between 1990 and 2020.[5] The second reason concerned politics because the firewall made Kyoto a political dead letter in the United States, where opponents labeled the exemption for developing countries, including China and other majors, as unfair. The famous Byrd-Hagel Resolution in July 1997, which declared that the United States should not sign onto a treaty committing developed countries to cut their greenhouse gas emissions unless it also called for specific commitments from developing countries, passed the Senate 95–0. So although the United States agreed to the final text in Kyoto, it was never able to join the agreement. Given the carbon footprint of the United States both in 1997 and historically as well as our power and influence, the Kyoto Protocol was effectively doomed from the start. In June 2001, the second president Bush unceremoniously took the United States out of the Kyoto process altogether. With the United States and all developing countries on the sidelines, the protocol covered less than 25 percent of global emissions.[6]

In light of that reality, countries at the 2007 COP in Bali, Indonesia, adopted the "Bali Action Plan," calling for a new agreement in 2009 covering all countries. That negotiation had been going on for a year when the Obama administration took office on January 20, 2009, and was supposed to conclude in Copenhagen less than eleven months later. Although no one knew it then, the fraught negotiation that lay ahead that year would become the first in a seven-year struggle that culminated in Paris.

* * *

When the Paris Agreement was adopted in 2015, it broke new ground. It was universal, casting aside the old paradigm in which obligations only applied to developed countries. It articulated strong goals to limit both global temperature and global greenhouse gas emissions. It established twin, staggered, five-year cycles so that it would be constantly renewed, with one cycle for individual countries to ratchet up their emission targets, and another to take stock of how the world in the aggregate is doing on reducing emissions in light of the latest science. It instituted a transparency system for countries to report on their progress and for those reports to be reviewed by international experts. The agreement differentiated between developed and developing countries in a new manner, not based on the old firewall. It also adopted a hybrid legal form, with nonbinding emission targets but binding procedural rules. And it called for the mobilization of finance to assist poor countries with their climate needs.

* * *

The years since the agreement was reached have been years of consequence. The United States and China, whose collaboration was pivotal to delivering the Paris Agreement, officially joined the agreement together on September 3 in Hangzhou, China, when Presidents Obama and Xi Jinping presented the relevant formal instruments to UN secretary-general Ban Ki-moon, and the agreement entered into force in near record time on November 4, thirty days after fifty-five countries accounting for at least 55 percent of global greenhouse gas emissions had officially joined. But four days later, on November 8, Donald Trump was elected president of the United States.

On June 1, 2017, Trump announced his plan to withdraw the United States from the agreement in a fact-free Rose Garden speech. The reaction around the world and at home was swift and emphatic. European leaders issued an immediate statement declaring that the Paris Agreement was irreversible. China made it clear that it would stay in the agreement with or without the United States. Within days, a new US coalition called We Are Still In circulated a declaration of support for the Paris Agreement signed by more than 3,900 CEOs, mayors, governors, and college presidents, among others. The next month, Governor Jerry Brown of California and former New York mayor Mike Bloomberg launched a new effort called America's Pledge to analyze and report on actions by states, cities, the private sector, and others to reduce emissions and bring the United States as close as possible to its 2025 Paris emissions target. In June, twenty-four state governors formed the US Climate Alliance, led by Brown, Jay Inslee (state of Washington) and Andrew Cuomo (New York).

The Intergovernmental Panel on Climate Change (IPCC), the United Nations' science body, continued to issue important reports underscoring the urgency of action, extreme weather events became still more extreme with each passing year, and greenhouse gas emissions kept rising despite the urgent need for them to peak and start declining. At the same time, the clean-tech story got better and better, with technology like wind and solar energy becoming dramatically cheaper and spreading dramatically faster than most anyone even dreamed of, much less predicted. Electric vehicles (EVs) and the batteries required both for those vehicles and to provide storage capacity for power plants using renewable energy started racing up the same exponential curve.[7] And a burgeoning world of innovation started booming to develop technologies that we need but don't yet have. Yet there continue to be obstacles in our path—powerful resistance from the global fossil fuel industry, complicated geopolitics, and the realities of human nature. Our capacity to overcome these hurdles quickly enough will make all the difference.

* * *

This book is about the struggle of the international community to negotiate the kind of agreement essential to meeting the challenge of climate change. Containing climate change requires action on many fronts. Nations, states, and cities need to enact laws and regulations aimed at spurring the shift to clean energy, protecting our "natural capital," and removing greenhouse gas emissions from agricultural production. The work of research scientists, innovative companies, and civil society is vitally important as well. But because climate change is a quintessentially global threat—the greenhouse gases released anywhere affect us everywhere—you can't contain it without sustained international engagement, and the Paris Agreement is the vehicle for that engagement. The book is also about the people and relationships that made the Paris Agreement happen, about negotiators and ministers who made a difference and often became my friends, from China, India, Brazil, Europe, Colombia, South Africa, New Zealand, the Marshall Islands, and all over the world. And it's about the practice of diplomacy in pursuit of a mission.

The agreement, of course, is far from perfect, and we are not yet collectively doing what is necessary to conquer the threat. In the final chapter of the book, I'll discuss some of the reasons for that as well as steps that can make the Paris regime work more effectively. But without Paris, the global effort to contain climate change would be adrift.

Barack Obama had it right on the day the Paris Agreement was adopted when he said that it "represents the best chance we have to save the one planet we've got."[8] He used the same motif not long after when he visited the State Department to thank my team for our work, signing a copy of the agreement's first page: "To the Paris Team—You've given future generations a fighting chance!" He had it right because there is no sure thing for a challenge like climate change, requiring a rapid, precedent-setting transformation of the global economy. Paris doesn't guarantee success because it cannot, but it gives us a structure, it gives us goals, it gives us rules, it makes demands, it brings us together, and it helps to build norms and expectations in the global public square. It gives us a chance.

1 Creative Destruction: The Start

As I joined my team in late January, I was, in effect, jumping onto a moving train since the 2007 Bali conference had launched a new round of climate negotiations for a major new treaty meant to conclude at the Copenhagen conference in late 2009, just ten months away. Global expectations for the new treaty were sky-high, in part because of the wave of confidence engendered by Barack Obama's election. But I knew that the paradigm for the new agreement envisioned by most countries and climate activists around the world—a legally binding agreement to reduce emissions, binding only on developed countries—would not work.

After all, I had lived through the Kyoto experience of watching a new climate agreement land dead on arrival in Washington. If a new Copenhagen agreement perpetuated a Kyoto-style firewall between developed and developing countries, it would again be politically toxic in the United States. Indeed, any kind of free pass for China—the world's largest emitter of greenhouse gases since 2005 and the world's second-largest economy—would itself doom US political support for a new agreement.

Mindful of these realities, I had two related objectives in mind as I stepped into my post: to avoid concluding another international climate agreement that the United States could not join and to chart a new path for effective global action, with the United States at the forefront of climate diplomacy.[1] The way to accomplish these objectives, as we saw it, was to land a meaningful new climate accord that started moving away from the firewall structure of Kyoto, aware that many developing countries saw that structure as not only essential but also their inherent right, and not only their right, but already embedded in what they saw as the foundational documents of climate diplomacy: the UNFCCC, Kyoto Protocol, and Bali Action Plan.

* * *

The foundation for everything we wanted to accomplish on the international front was the rebuilding of US credibility after a Bush administration that disengaged from the Kyoto process and did little to cut US emissions at home. President Obama called for strong climate action while still on the campaign trail and started to deliver as soon as he took office. He provided more than $90 billion in strategic clean energy investments and tax incentives as part of his roughly $800 billion stimulus package to rescue the staggering US economy;[2] he announced major plans to increase mileage standards for cars and light trucks in 2009[3] and boosted those standards in 2011 to a doubling of miles per gallon from approximately twenty-seven to fifty-four miles per gallon by 2025;[4] and he pushed signature "cap-and-trade" legislation to begin cutting greenhouse gas emissions right away and reach 80 percent below 2005 levels by 2050.[5] All of these actions demonstrated that Obama meant business, and they put wind at our backs on the international stage.

Moreover, on her first trip as secretary of state—to China, Japan, Korea, and Indonesia—Secretary Clinton signaled that global climate change was a pillar of her three-part agenda, along with the global economic crisis and security challenges such as Afghanistan and North Korea. And she brought me along, placed me prominently in her delegation, and introduced me to every leader, underscoring the importance of climate for the Obama administration and making clear that I was the United States' international climate lead.

* * *

Back in Washington after the trip, my team and I spent the next several months developing a US approach on key issues for the Copenhagen negotiations, hosting a number of important diplomatic visitors, and launching a new diplomatic vehicle. In early March, Sue, Jonathan, and I had dinner with a team led by the United Kingdom's lead negotiator, Pete Betts, whom I first met at the Buenos Aires COP in 1998 during my climate days in the Clinton White House. I always liked Pete. He was smart, strategic, and blunt, and always knew what was up in the European Union as well as the United Kingdom. He didn't mince words, never sugarcoated anything, and had little patience for diplomatic niceties. All through my time at State, I sought him out, learned things from him, and always knew where he stood.

He also understood our political constraints, and in the spirit of blue-sky, what-if thinking, wanted to use our dinner to explore the idea of countries submitting their own national plans as part of a new agreement rather than a specific emission reduction target. Pete was clearly searching for a way forward that might work for the "Yanks," as he always called us, guessing it would be easier for us to get an international agreement through the minefield of Senate approval if we were simply agreeing to do internationally what Congress had already agreed to do domestically. I saw this as an example of the "special relationship" people always attribute to the United States and United Kingdom, which I found, most of the time, to be quite real. The UK, at that time, was still squarely situated within the European Union, yet their diplomats often had a better feel than their continental partners for how to work effectively with the United States. And that same understanding extended to Pete's boss, Ed Miliband, the UK's energy and climate change minister, whom I would meet in April.

Sue and I came to refer to the UK idea of inscribing national policies in a new climate deal as the "Agraria" proposal, after the waterfront restaurant where the Brits had suggested it. It turned out that the Australians had a related idea: that all countries submit "schedules" of actions they proposed to take to reduce greenhouse gas emissions, borrowing a term from the world of trade. And even before meeting with Pete and his team, Jonathan, Sue, and I had started thinking along similar lines at a coffee we had together during the Obama transition. All of these related ideas were potential alternatives to the unworkable Kyoto model in which developed countries alone took negotiated emission targets.

I also met Denmark's environment minister, Connie Hedegaard, during this time. She was intense, passionate, serious, and determined to get things done. Unlike Pete, she was not focused on considering a new way forward. She knew what the world needed, she was a woman on a mission, and her purpose was to persuade. As we sat across from each other at a narrow table in a State Department conference room, Connie leaned over on both elbows, eyes on me, pressing her case and addressing me vigorously as "Todd Stern." The United States needed to do much more, she said, the world needed a legally binding agreement, and so on. Connie turned out to be an important player on the international climate stage for years, since less than two months after the Copenhagen conference, she was named the European Union's first commissioner for climate action.

On March 16, I met in Washington with my Chinese counterpart, Xie Zhenhua. Given the sheer size of China's rapidly growing carbon footprint and influence in the developing world, no country was more important to the mission of arresting climate change. Xie was the vice chair of the powerful National Development and Reform Commission (NDRC), which was responsible for the preparation of China's famous five-year plans, among many other things. He was China's climate minister, responsible for both climate policy at home and climate negotiations abroad. I had not seen him during Secretary Clinton's trip in February since he had been traveling, so our March session was the first of what would become scores of meetings on the seven-year road to the Paris Agreement. We spoke through interpreters.

Xie has a round, open, expressive face that conveys a broad range of attitudes—concentrated, sincere, surprised, mocking, delighted, sarcastic. He gestures broadly in the midst of negotiating sessions, as if to say, "How can you possibly disagree with me," or "Come on, are you kidding?" He smiles broadly, laughs easily, and when he gets angry, his face turns a deep shade of red and he jabs the air in a sharp staccato. He is smart, well-versed on the issues, clever, wily, and every inch a leader. If you start to box him in on an important point, his go-to maneuver is to attack, often on an unrelated issue, on the theory (which he admitted once) that the best defense is a good offense. He has a habit, after a long back and forth on a contested issue, of presenting his summary, and that's the moment you need to keep your hand on your wallet because if you're not careful, you'll discover that your pocket has been picked. Over the years, we argued, reasoned, wrestled, laughed, persevered. We became good friends.

At that first March meeting, I made clear to Minister Xie that President Obama was deeply committed to climate action at home and abroad, but that we wouldn't accept the Kyoto structure in a new agreement. I acknowledged China's enormous economic and development challenges as well as the progress it was already making on clean energy, and said I thought we could collaborate together productively. In particular, I said I thought a close working climate relationship between our countries could become a positive pillar in our broader bilateral relationship, which was typically marked by stresses and strains. This point became a strategic foundation for the US-China climate relationship during the Obama years—a climate relationship that had been historically acrimonious. Xie seemed open to my suggestion, and over time it became a shared watchword. It didn't prevent us from fighting tooth and nail over concepts, words, and nuances all the way to Paris, but in time it became an important anchor in the way both sides conducted themselves. Nearly seven years later, on the first day of the Paris conference in 2015, when President Xi met with President Obama and told him that he saw climate change as a bright spot in our relationship, I thought about how the early seed I planted had born fruit, with watering by a lot of people, all the way up to the two secretaries of state I served, the White House climate team, and the president.

Xie asked me at our March meeting what I thought China should do on climate change. I didn't answer him in detail then, but I did when we next met in China in June. At that time, I ticked through an ambitious list. I said that we'd like to see a long-term pathway of emission reductions to 2050 with Chinese emissions peaking in the 2020–2025 period, midterm steps to reduce emissions, and an internationally binding commitment to carry them out. I included this last point not because we were keen to have a legally binding agreement in Copenhagen but instead to make it clear that if the agreement were to be legally binding, it would have to bind China as well. I added that China should be subject to a transparency system of reporting and review of its climate performance. I said that without measures like these, it would be hard to imagine US Senate approval of an international agreement. Xie responded that my "requirements" were "completely impractical," we couldn't expect China to peak its emissions only ten years after the United States, China wouldn't consider binding targets until it had finished industrializing, and any transparency measures were subject to the Framework Convention principle of CBDR, by which he meant the firewall.[6] For good measure, he noted that whether China achieved its emission targets would depend on whether developed countries provided assistance—a comment I took more as a means of China insisting on its developing country status than as an actual demand for money.

Near the end of March, I traveled to Bonn for my first appearance at a multilateral forum, one of the so-called intersessional meetings of all countries held every year to prepare for the year-end COP. These meetings, usually two or three per year, are negotiator-level sessions, a notch below ministers, so I normally didn't attend. But for this first one of the Obama era, I wanted to send a message by showing up. When it was my turn to speak in the large, UN-size hall of the old Hotel Maritime, I said that on behalf of President Obama, the United States was back, wanted to make up for lost time, and was seized with the urgency of the task before us. I said that we needed to be guided by science and pragmatism, and that the United States recognized our unique responsibility as the world's largest historic emitter, and then hit the high points of President Obama's aggressive climate program at home. I said that the responsibilities of developed and developing countries could be differentiated in a variety of ways, but that "the unforgiving math of accumulating emissions" meant that all key countries had to take strong action that was measurable and verifiable. I acknowledged that poor countries needed assistance to develop in a sustainable way and have the capacity to adapt to the impacts of climate change that were already happening. I ended by underscoring the shared potential we had if we all did what we could. The hall erupted in sustained applause, delighted to turn the page on the Bush administration and hear an activist message from a new envoy for a new president. I turned to Jonathan and Sue, whispering, "Enjoy it now, because that will be the last ovation we ever get." I knew that before long, we would have to take unpopular positions in order to land a workable outcome in Copenhagen.

After Bonn, Sue and I went to work preparing a read-ahead memo for a "Principals Committee" meeting at the White House with the relevant cabinet secretaries and senior White House aides. We wanted to socialize our thinking about the kind of agreement we favored for Copenhagen and get guidance from the committee. We outlined two different options. In option 1, more conventional, developed countries would agree to binding emission targets while developing countries would submit their relevant domestic laws and regulations to a "registry." Option 2 was more akin to the Australian schedules idea, calling on all countries to put forth and implement domestic laws and regulations. We met with the principals in late April in the Old Executive Office Building with around twenty cabinet secretaries or deputies and a few senior White House officials arrayed around a long table. On balance, the committee thought the downside risk of publicly breaking away from the more conventional approach so early in Obama's presidency was too great given that most countries and the environmental community assumed an outcome roughly like option 1. Sue and I were convinced that we would eventually end up with an approach closer to option 2, but I thought the committee's decision to be more cautious early on made sense.

* * *

In March, my team and I also worked with the White House on a new diplomatic vehicle that reshaped an initiative the Bush administration started in May 2007 for a group of seventeen major developed and developing economies called the Major Economies Meeting on Energy Security and Climate Change (the MEM).[7] This was a good idea, and indeed I had proposed something similar in an article I published in January 2007 with a Clinton era White House colleague, Bill Antholis, calling for the establishment of an "E8" of developed and developing countries to focus on climate change and other existential environmental threats.[8] (I assume our article had nothing to do with the Bush effort.) Having witnessed the cacophony of the COPs in Kyoto (1997) and Buenos Aires (1998), where negotiators from over 190 countries struggled to manage too many meetings in too little time with not enough engagement by country ministers, I concluded that the process would be well served by a small, high-level forum where ministers, together with top negotiators, could meet regularly in a quiet, collegial setting to exchange views about what needed to be done and how. The Bush administration initiative included the right countries, but the group lacked a real mission, plus the Bush team had no claim to climate leadership given its half-hearted domestic effort, the way it abandoned the Kyoto process without even an effort to amend it, and the embarrassingly inadequate climate goal it announced in April 2008—no emission reduction at all, just a peaking of emissions by 2025.[9]

We stuck with the same group of countries, rechristened the group with a revised name to steer clear of the Bush administration's low-octane climate brand, and infused it with the core mission of making real progress in the climate negotiations. Since 2009 was President Obama's first year and the stakes were high for Copenhagen, we decided to hold a leaders-level meeting of the new Major Economies Forum on Energy and Climate (MEF) as part of the July "G8+5" gathering in Italy. We held three preparatory sessions in April, May, and June for the leaders' representatives—ministers or senior advisers—plus a last-minute session in Rome two days before the leaders' meeting to settle the last remaining issues in the Leaders' Declaration.[10]

The MEF proved to be one of our most important diplomatic initiatives, appreciated by all the participants. Since the United States set the agenda, the MEF allowed us to decide what issues to focus on several times a year, and we deliberately teed up the toughest ones, where focused dialogue could make a difference. We chaired six meetings in 2009, and generally three or four meetings per year after that. Meetings usually lasted a day and a half, including a dinner. We typically invited a few smaller, but influential developing countries as well as the executive secretary of the UNFCCC and the host country of that year's COP. Other than in the special case of the 2009 leaders' meeting, we never tried to produce what governments call "deliverables," such as a negotiated declaration. We wanted to take the pressure off participants to produce an outcome, instead encouraging an open atmosphere of candid conversation and debate where people felt free to road test ideas that might bridge differences. And we paid close attention to creating a cordial atmosphere, keeping the size of country delegations limited (usually to four), paying attention to the size and shape of the table to keep the feeling intimate, and making sure dinners had good acoustics and good food. I recruited a smart, tough, skinny veteran of G7 and G20 meetings with Bob Dylan hair, Paul Brown, to help get us started and soon thereafter the unflappable veteran Isabel Gates to run all the important logistics.

Mike Froman, the deputy national security adviser for international economics and my principal White House counterpart, chaired the meetings while I represented the United States. Mike took pains to act as a fair arbiter rather than a proponent for our side. Other participants appreciated him for his manner, intelligence, and humor, and the direct imprimatur of the White House that he brought into the room. Mike and I had an easy, trusting relationship and made a good team both in the context of the MEF and more broadly. He is smart, funny, and unflappable, has great judgment, and was someone from whom I could always get sound advice. He also had clout inside the White House as a skilled and well-liked player with a key portfolio who also happened to have been a classmate and friend of Barack Obama at Harvard Law School.

We hosted the first MEF meeting April 26–28 at the State Department, and it got us off to a good start. In the morning, Clinton made opening remarks, at the end of the day Obama welcomed the slightly dazzled attendees to an unannounced get together in the White House Blue Room, and in the evening we gathered for a reception at the Kennedy Center. During the meeting itself, we opened the door for ministers to say whatever was on their mind since the real purpose of this first encounter was to establish a foundation of trust and good feeling; a more prescriptive agenda of the kind we used in subsequent meetings could wait. But as an important signal of what substantive MEF discussions could be, South African minister Marthinus van Schalkwyk and I found ourselves in an impromptu debate on the second day of the meeting. Marthinus was strong, confident, and effective with the bearing of a leader. Seated near me, he took the floor to chastise the United States for proposing a 2020 emission reduction target he considered too weak. In fact, we were months away from deciding on an actual Copenhagen target, but the president's budget, submitted in February, included a legislative proposal to reduce emissions 14 percent below 2005 levels by 2020, escalating to an 83 percent reduction by 2050. In view of what our emissions were in 2005 as well as our 2050 number, this was a respectable proposal, but Marthinus argued that compared to a 1990 baseline, the benchmark under Kyoto, our 2020 number was weak, and he couldn't sell it to South Africans.

I had heard the same refrain from Europeans on a number of occasions and didn't buy it.[11] We were seeking comprehensive legislation, which if enacted, would get the United States off to a good start cutting emissions and escalate rapidly to a huge reduction by 2050. So I answered Marthinus, defending the target reflected in our budget and providing some background about the political realities of securing strong climate legislation in the United States. Marthinus was not satisfied, and he and I then banged the ball back and forth across the net for several minutes, neither giving ground. Mike, as chair, let it happen, temporarily ignoring the list of those who had raised their placards to speak. This might not seem a big deal but it was, compared to the typical practice in multilateral meetings where country reps raise their flags, read their talking points, and rarely interact. The exchange between Marthinus and me made a statement. It said that the MEF was going to be different. Here we would debate, argue, and engage. Over the years, no one was a bigger booster of the MEF than Minister Xie, and it was precisely this quality, together with the US penchant for putting the trickiest issues on the table, that most attracted him.

The next two scheduled meetings, in Paris (May 25–26) and Cuernavaca, Mexico (June 22–23), were aimed at negotiating the Leaders' Declaration for the July 9 meeting in L'Aquila. These negotiations were a struggle all the way until the end of an unscheduled extra session in Rome, two days before the leaders' meeting. The Europeans argued for recognizing a science-based goal of holding global temperature rise to 2 degrees Celsius (2°C) above preindustrial levels, and to meet that goal, called for aggregate emission cuts of 80 percent for developed countries by 2050, compared to 1990, and

50 percent for the whole world. The United States focused more on creating as much parallelism as possible between what developed and the major developing countries were called on to do in limiting their emissions as well as subjecting those mitigation efforts to measurement, reporting, and verification (MRV). China and India, with support from Brazil and South Africa, resisted the European and US arguments. They wanted to preserve the firewall, and China, in particular, wanted to avoid long-term aggregate commitments that it feared would constrain its capacity to grow.

Before the start of that last session in Rome, Xie pulled me into a small side room, anxious to tell me China could not tolerate language calling for a 50 percent global reduction in emissions by 2050 because it had calculated that such a goal would impose infeasible demands on it. I decided on the spot not to fight about this. I knew that China would not move on this issue at that point and thought it wiser for me to show that I understood its constraints, so I told Xie that I would not push him. This was a debatable call, but one that proved an important moment in building our relationship. It was a sign to him that I would at least try to understand China's needs. On a number of occasions in subsequent years, he cited that brief conversation as a kind of emblem of our special relationship.

In the end, the declaration served as a useful step forward. It included a soft version of the 2°C goal, but one that strengthened over time, illustrating a common feature of negotiations—that negotiators, if they can't do better, will often settle for establishing an initial foothold on an issue, and then in due course, try to climb higher. The declaration also called on all MEF countries, not just some, to take prompt action to limit their emissions, and stressed that such action needed to be transparent and subject to "applicable" MRV. Moreover, it included a reference to the Mexican idea of a Green Fund to be funded by all nations, developed and developing, and provide financial assistance to countries in need.[12] And it led to the creation of a new entity called the Clean Energy Ministerial, proposed by the United States and headed by energy ministers to explore ways that MEF countries—accounting for roughly 80 percent of global emissions—could pursue clean energy progress together.[13]

The actual leaders' discussion broke no ground, but was valuable as a means of elevating climate change to a discussion among this group of leading presidents and prime ministers. President Obama argued that while we can't expect developing countries to take the same actions as developed, they needed to take strong action given their rapid emissions growth. Gordon Brown, UK prime minister, pressed for concrete, long-term emission goals and ramping up climate finance for developing countries to $100 billion per year by 2020, with much of that coming from the private sector. Australian prime minister Kevin Rudd urged a "grand bargain" between developed and developing countries, since developed countries had to recognize their historical responsibility, but developing countries would be the dominant emitters going forward. After the leaders' session, Mike and I walked with the president back toward his quarters at Roma House in the military barracks where this G8 was staged.[14] He was in a reflective mood, recognizing the urgent need for strong action by all of these countries, but also the difficulties for those, such as Indian prime minister Manmohan Singh, who face such fundamental challenges regarding development and poverty. He was frustrated at how difficult it was to make more rapid global progress.

I expect the lesson that most countries took away from the difficult negotiation of the Leaders' Declaration was that progress toward a Copenhagen agreement would come hard. The declaration, after all, was a short, high-level document among just 17 parties—a far cry from a full-blown agreement among more than 190. But the process of negotiating it was constructive. While we took care never to treat the MEF as an actual negotiating forum since it was not a UNFCCC body and included only a small number of countries, negotiating the declaration enabled MEF ministers to get accustomed to testing ideas, seeking solutions, and wrestling with language. All of that contributed to the MEF becoming an effective and important forum.

Denmark, president of COP 15 and represented in our meetings by Prime Minister Lars Løkke Rasmussen's adviser on national security and climate change, Bo Lidegaard, took away two lessons of its own. Bo saw that although difficult, it was possible to find textual language on core issues that all major players could accept—a departure from the sclerotic and disputatious process that played out in the formal UNFCCC intersessional meetings, where the compilation text had ballooned to nearly two hundred pages of bracketed, conflicting options. And he concluded that Denmark would have to invite leaders to the COP—something never done before—because only the pressure of leaders attending the MEF meeting in L'Aquila made it possible to secure even the modest Leaders' Declaration.

* * *

A week after the MEF leaders' meeting, in the last important diplomatic event of the summer, I joined Secretary Clinton on a four-day trip to Mumbai and New Delhi, which proved most significant for me as the time and place I met India's environment and forestry minister, Jairam Ramesh. India was a crucial player in the negotiations historically and continued to be. It was one of the world's largest emitters of greenhouse gases—though its emissions were less than a quarter the size of China's. It was influential among developing countries. And dating back to Indira Gandhi's appearance at the Stockholm Conference on the Human Environment in 1972, it had a history of insisting on the primacy of development over environment.[15] Moreover, in 2009, India was one of the rare countries outside the United States where climate change was a highly charged political issue, focused, in India's case, on perceived unfair pressure from rich countries, which the Indian press avidly and at times sensationally covered. I saw this firsthand on our arrival, when a page three headline in the *Times of India* blared, "Climate Man's Visit Shocks India." This made me laugh, but also made me sensitive to the political backdrop against which Indian negotiators worked.

Secretary Clinton and I met Ramesh at the ITC Green Center in Gurgaon, a midsize city on the outskirts of New Delhi. Ramesh was an interesting character—highly intelligent, agile, politically savvy, and charismatic, with long swept-back hair, a sartorial flair in traditional Indian attire, and a ready sense of humor. He was also well-connected to Sonia Gandhi, leader of the then dominant Congress Party and a member of India's most famous political family. Up until this trip to New Delhi, I had dealt with Prime Minister Singh's climate adviser, Shyam Saran, a formal, dour, bald, often perturbed-looking old-school traditionalist trained in the Ministry of External Affairs and given to short, sharp, irritable lectures. Saran had taken positions in our various MEF meetings much as I had expected—as a stalwart defender of the traditional firewall. Ramesh, I would discover, was cut from an entirely different cloth. Still, our first interaction with him was not auspicious since he more or less sandbagged Secretary Clinton to get a headline. The ITC event included a closed-door roundtable with business and environmental notables and a press conference. At the roundtable, Ramesh, who was never less than gracious and charming when alone with the secretary, announced defiantly that India would not agree to a legally binding commitment to limit its emissions (we knew that and weren't pressing them) and complained about US pressure: "There is simply no case for the pressure that we, who have among the lowest emissions per capita, face to actually reduce emissions."[16]

At the news conference, Ramesh handed out copies of the written account of what he had said in the closed-door session. This stunt, especially the printed copies of his remarks, annoyed me, but Clinton, who rolled her eyes, didn't give it much thought. I came to understand that Ramesh did this kind of thing quite mindfully to give himself room to maneuver on the road to finding solutions. He was a freethinker, untroubled about flouting Indian orthodoxy, and a believer in the need for climate action. He also thought that breaking away from old doctrine would be better for India's own standing and reputation. And he was fearless in pressing forward, sometimes getting dragged back, whether by others in India or China, but in short order looking again for a new way around, like an incorrigible kid who won't be set straight.

2 Creative Destruction: The Pivot

On September 4, with just three months left before the start of the Copenhagen conference (COP 15) and little perceptible progress in the formal negotiations, Bo Lidegaard, the Danish prime minister's chief climate strategist, came to see us in Washington. Bo was tall, blond, restless, charismatic, determined to find a way forward, and well-versed in climate change, going back to the days of the Framework Convention. He was also a historian, author, broad-gauged thinker eager to talk about anything, high-octane activist, and gambler with nerve. He had an idea he wanted to discuss.

Bo had first started talking to Sue and me about a potential new approach toward the end of the June MEF meeting on a secluded patio in Cuernavaca, but his thinking had advanced. Over lunch with Sue, Jonathan, and me at Kincaid, our restaurant of choice near the State Department, Bo described a plan that marked a sharp departure from climate orthodoxy. He envisioned a short, operational agreement of around six to eight pages that would be politically rather than legally binding. Developed and developing countries would both commit at the leader level to implement their own national targets or actions. The agreement would include reporting and review provisions to ensure that all were taking action. Developed countries would agree to provide financial and technology assistance to developing countries. And major countries, developed and developing, would have to put forward their mitigation targets or actions before the COP.

The Danish plan abandoned the precept of legal bindingness because Bo believed there wasn't enough time to complete a full legal agreement in Copenhagen, and more fundamentally, that an agreement requiring US Senate approval would be doomed.

Right away we saw the Danish plan as a potential game changer. It was consistent with ideas Sue, Jonathan, and I had long discussed, most recently in an August memo for the White House. Moreover, that it came from Denmark rather than the United States was highly desirable from a tactical standpoint both because Denmark, holding the COP presidency, would be in a good position to socialize the idea and because it was better for an idea like this *not* to come from the United States, given the flack it would attract from multiple sources, including developing countries, the European Union, and the environmental community. However pleased countries were to have the United States back in the game, the irritations and resentments of dealing with American power and influence had not disappeared. The only snag for us involved Bo's last requirement. We told him that we could not announce a US target before the Waxman-Markey cap-and-trade legislation was done since that would risk jeopardizing the legislation itself. But Bo insisted the plan could only work with US targets on the table in Copenhagen. Shortly after lunch, Sue started musing about ways to work around this obstacle, such as by announcing a target before the COP, but deeming it provisional until our legislation was enacted.

In Europe, though, the reaction to the new Danish plan was skeptical, if not hostile. The European Union, Germany, UNFCCC secretariat, and even Environment Minister Hedegaard's side of the Danish climate team were convinced that the only way we could combat climate change effectively was with a legally binding treaty that included emission reduction targets keyed to the science-based goal of holding the increase in global average temperatures to below 2°C. In their view, anything less would be a failure. They knew that most developing countries would not accept legally binding obligations for themselves, but argued that those countries would do so eventually if developed countries led by example. By this logic, the United States was the linchpin to making the European vision work. If the United States would just say yes, a legally binding agreement in Copenhagen could be reached, and China and others would join the binding regime in due course. Letting the United States off the hook now, in 2009, would be a fatal mistake in their view. The right response to the bogged down negotiations was to ramp up pressure on the US government. Obama, they reasoned, would have to come on board, and if he didn't agree this year, he would soon. If the negotiations had to go beyond 2009 to get this done right, so be it. They saw no reason to betray the goal they had pursued for years in order to land a cheap prize in Copenhagen.

But Bo and Prime Minister Rasmussen disagreed. They thought securing an actual agreement in Copenhagen, even if a limited one, was essential. They contended that the best way forward was to capture the real progress we could make now—with countries promptly adopting real targets and plans to reduce emissions, putting a transparency (MRV) regime in place, agreeing on stronger financial and technology assistance to poor countries, and resuming their effort to negotiate a full legally binding agreement after the Copenhagen COP.[1]

We thought Bo and his boss were right. Legally binding for us now and for China at some point down the road would never fly politically in the United States. Moreover, I never bought the European claim that countries like China and India would voluntarily accept binding targets in the near future simply because they saw developed countries go first. My own hard-headed view was that when you put something in somebody's pocket—in this case, special, lenient treatment behind a firewall—it tends to stay there until someone applies enough pressure to dislodge it, and that is never easy to do. Kyoto put the firewall in developing country pockets in 1997 with no time limit or phaseout. Now we were negotiating a new agreement with a chance to start resetting that arrangement, and I thought we had to make the most of the opportunity.

I told the president about the Danish idea at a September 16 meeting in the Oval Office with senior White House advisers, mainly focused on the UN climate change speech he would make at the upcoming annual UN General Assembly meetings. He agreed that the Danish idea was promising and we should work to advance it, though not yet publicly. He also agreed with us that a focus on major developing countries was the right way to go since it would be a bridge too far to insist that all developing countries do these things. We thought that if the developing countries in the MEF were on board to put forward their own targets, we would be in business.

China

Minister Xie and I spent a lot of time together that fall, in Washington, New York, London, Beijing, and Copenhagen, and our relationship was growing more and more congenial, even if we were still far apart on substance. This was evident at the start of our September bilateral meeting in Washington. As we greeted each other in my State Department conference room, I said that I had a funny story to tell him. My wife Jen and I were driving one day in the summer with our nine-year-old, Zachary, and four-year-old, Ben, in the back seat. It was around noon, and the boys were hungry, arguing about where we should have lunch. Zachary wanted to go to Panera Bread, but Ben didn't. This went on and on. Ben would not budge. Finally, an exasperated Zachary gave in, but said that meant we'd order from Surfside, his choice, for dinner. Ben answered, "I agree about lunch, but I don't agree about dinner." I turned to Jen and said, "Ben must be getting negotiating lessons from Mr. Xie." When I finished the story, Xie and his whole team roared, and Xie, bright red and laughing, said in Chinese, "No, no, he learn from his daddy!" So we were definitely getting to be friends, but that didn't yet help on the issues. This was aggravating but not surprising. Negotiators, after all, act in what they perceive as their country's national interests, whether they like each other or not. Still, relationships matter, trusting your counterpart matters, and the rapport that Minister Xie and I were developing would prove valuable over time. For now, though, the Chinese were not budging. And they would prove to be the biggest stumbling block for any shift away from the old firewall paradigm.

At a bilateral meeting in New York City in September, Xie said any suggestion of "committing" to actions was a redline for China whether the commitment was called "legal" or "political," and he added that it would be a redline to say that a developing country pledge to take action was supposed to have the same force as a developed country pledge. I also started to notice in our September and October meetings, as MEF countries discussed the importance of submitting their own pledges to the UNFCCC, Minister Xie kept talking about how China would issue a press release or white paper on what they planned to do. This was no accident. China was trying to avoid any implication that it would submit itself to the authority of the UNFCCC. Xie often indicated that we would find a way to handle this last issue, and that kept me off-balance, thinking he might show flexibility in the end. To underscore the real difference between a formal submission to the UNFCCC secretariat and a unilateral announcement, I started talking in the MEF and elsewhere about the importance of "internationalizing" our targets. I learned much later that Bo was focused on this issue too, struggling to persuade India and China to "cross the bridge" that would take them from announcing purely domestic actions to making international political commitments.

I also struggled with Xie over the verification, or review, element of MRV. Xie made it clear that on that issue, he agreed with India's Shyam Saran, Prime Minister Singh's climate adviser. At our September 17–18 MEF meeting in Washington, Saran argued that there could be international review of developing country actions only if those actions were financed by developed countries—"supported" in UNFCCC lingo. This meant no review for most developing country action and virtually none for China, since it financed the vast majority of its own actions. For all of those "unsupported" actions, Saran and Xie contended that transparency should be limited to self-reporting. This argument was a nonstarter for us as well as for the European Union, led by the tough-minded Karl Falkenberg, an astute negotiator trained in the world of trade. After all, transparency allows country performance to be tracked. And this issue mattered to us politically as well as substantively. In my first meeting on Capitol Hill, in February, with Senator Kerry, then the Senate's conscience on climate change, he hammered this point, saying that unless China's climate actions were both reported and verified, members of Congress would never trust what they claimed they were doing.

Ramesh Steps Up

On September 20, at a so-called Greenland Dialogue meeting of around fifty countries that Danish minister Hedegaard chaired at the Arrowwood Conference Center in Westchester, New York, India's Jairam Ramesh took the floor and broke ranks with both Xie and his own Indian colleague Saran. Ramesh suggested that the kind of international reviews of both developed and developing country behavior carried out by the International Monetary Fund (IMF) and World Trade Organization (WTO) might serve as interesting precedents for the UNFCCC. This was radical stuff. Ramesh was abandoning Saran and Xie's notion that only the supported actions of developing countries should be subject to reporting and review. He was citing two major, frontline international organizations where transparency measures were used for all countries, without fuss, effectively throwing down the gauntlet and saying, "Why not us?" And he was doing this as the representative not only of a major developing country but also one with a long history of allegiance to the firewall. In the IMF, member countries are required to participate in regular Article IV consultations, during which a team of economists from the IMF visits the country to assess economic and financial developments as well as discuss the country's relevant policies with government and central bank officials. WTO member countries are also subject to international scrutiny during periodic Trade Policy Reviews that include a policy statement by the country under review, and a detailed report written by the WTO secretariat. In 2009, when Ramesh spoke, the four largest trading countries, including China, were subject to a Trade Policy Review every two years, the next sixteen largest every four years, and all other countries every six years. Thus Ramesh, the inveterate freethinker, was suggesting a robust form of international review as a model for developing countries in the climate context, untroubled by the hard-line positions Saran and Xie had staked out just a few days earlier. And he was just getting started.

Obama Engages

On October 1, President Obama traveled to Copenhagen mostly to try to persuade the International Olympic Committee to select Chicago as the host city for the 2016 Summer Olympics (no luck there), but also, on my advice, for a meeting with Danish prime minister Rasmussen. During the meeting with Rasmussen, which Bo and I attended, the president conveyed broad support for the Danish concept of a Copenhagen Accord. Rasmussen stressed that to close a deal in Copenhagen, the heads of government, including Obama, had to come because only leaders would have the requisite political confidence to accept the kind of departure from the norm that the Danes were proposing. Obama cautioned against bringing heads to Copenhagen if a deal was not ripe—famous last words—but made it clear that he was prepared to attend if the circumstances were right.

On October 7 and 14, the president attended two internal meetings in the Roosevelt Room focused mostly on securing his approval of a US 2020 target based on the one embedded in the Waxman-Markey cap-and-trade legislation, whose passage by the House of Representatives in June was a big step, noticed around the world. Waxman-Markey called for a 17 percent reduction of greenhouse gas emissions below 2005 levels by 2020 on the way to a reduction of over 80 percent by 2050. Larry Summers, my old boss at Treasury in 1999–2000, inevitably rumpled and untucked, but dauntingly smart and wickedly clever, played the role of Cassandra, warning that the cost of emission "allowances" under Waxman-Markey would have to rise to uncomfortably high levels to hit a reduction target as ambitious as 17 percent. But Obama, unfazed by Larry's warning, approved the 17 percent target. And I flashed back, déjà vu style, to another presidential meeting I attended twelve years earlier in the Oval Office with President Clinton and his economic and environmental teams. They were there to discuss the emission reduction target the United States should bring to the upcoming Kyoto conference. The economists went first and laid out all the sober, pessimistic reasons why the environmental team's recommendations were impractical, unaffordable, and unwise. President Clinton reflected for a moment, looked up at the economic team, and said, "Why, y'all are just a bunch of lemon suckers."

The Sprint to Copenhagen

In the seven weeks between the end of a London MEF meeting in October and start of the Copenhagen conference on December 7, the Danes worked assiduously to build support for their proposal and so did we. Among other things, Sue and I went to Copenhagen four times before the COP began for meetings hosted by the Danes. I also took a trip to Shanghai for an extended session with Xie and another to Barcelona for a Greenland Dialogue meeting hosted by Connie Hedegaard. In addition, I met in Washington with the heads of top environmental groups and with Ethiopian prime minister Meles Zenawi, the most important African leader on climate change in 2009.

On October 20 right after the London MEF meeting, Sue and I met with Bo and his team in the prime minister's office together with several Australian negotiators and first saw a draft Danish text.[2] We thought it went overboard trying to placate the firewall demands of developing countries, but it was a start. Bo also told us that Denmark had discussed its idea with Mexican and Indonesian negotiators as well as the leaders of India and Brazil, all of whom seemed supportive. Xie, however, told Bo that Denmark should stick to the formal negotiation process, that a short-form political agreement would be a fatal mistake, and that Denmark should not go to leaders over the heads of negotiators.

Bo convened his next meeting on the Danish proposal November 7–8 in a beautiful, 260-year-old former warehouse on the Copenhagen harbor called

Eigtveds Pakhus, now run by the Foreign Ministry. The European Union, United Kingdom, France, Germany, Japan, Australia, and the United States attended, and the most important takeaway was that however unhappy our European friends were with Denmark's proposal, they were prepared to roll up their sleeves and try to get it done.

On November 11, I sat down with Prime Minister Meles to solicit his support for the Danish plan. A former revolutionary leader in Ethiopia, Meles was tough and shrewd, and in August had been appointed chair of the African Heads of State and Government on Climate Change, one of several influential African groups that played a role on climate. Meles was comfortable with the Danish idea of a short, political agreement, to be followed as soon as practicable by a full treaty, but he had four conditions: that the agreement be genuinely substantive, not just a declaration; that it clearly say the rise in global average temperature must not exceed 2°C; that there be greater focus than in the past on funding for adaptation (steps countries can take to build resilience and manage the impact of extreme events), with funding directed to the poor and vulnerable, rather than countries like China and India; and that the developed countries agree to provide developing countries $100 billion per year by 2020, the target first proposed by the United Kingdom's Gordon Brown. Meles didn't worry about going against the grain of China or others, having faced risks in his life a lot more serious than aggravating China over language in a climate agreement. He knew what he wanted and was prepared to deal.

Denmark hosted its "pre-COP" meeting for around sixty countries on November 17–18 in Copenhagen. For us, the most important part occurred before the meeting even started, during an early morning walk I took with India's Ramesh around Copenhagen's set of human-made rectangular lakes that passed right in front of the Scandic Hotel, where the pre-COP was held and where we were both staying. The setting was both picturesque and private. Among other things, we talked about the still unsettled dispute over transparency—in particular, the US stance that there had to be some form of international review of reports developing countries would prepare about both their emission inventories and the progress they had made in implementing their pledges.[3] Ramesh thought that we could accomplish what we wanted if we avoided certain buzzwords that developing countries regarded as applying only to developed countries, such as "MRV" and "verification," and that the review process for unsupported actions could be robust, as it is in institutions like the IMF and WTO. He advised using a term such as "consultative process" rather than "review." I told him I thought this kind of approach might work if we could get the words right and the relevant developing countries would agree. Ramesh talked about his idea during the formal pre-COP session, though he pulled in his horns by omitting reference to the IMF or WTO precedents—an omission that significantly diminished the impact of what he was proposing. But diminished or not, the Chinese got the point and were not happy. During one of the breaks in the action, Xie and his deputy Su Wei could be seen huddled with Ramesh in a discussion that looked a lot like taking Ramesh to the woodshed. After that break, Ramesh did not stick his neck out for a while. But no one kept Ramesh down for long.

On the second day of the meeting, Prime Minister Rasmussen laid out his case for the Danish plan more vigorously and clearly than he had ever done. But opposition to the plan was still strong. Minister Xie asserted that the plan would dilute the principle of CBDR and argued for an agreement that was legally binding for developed countries only. The Sudanese negotiator with the arresting name of Lumumba Stanislaus-Kaw Di-Aping, who also served as chair of the G77 group of nearly all developing countries, flatly rejected the Danish idea, as did Saudi Arabia's exuberantly obstructionist negotiator Mohammed Al Sabban, who accused developed countries of trying to change the terms of reference of the Bali Action Plan, renegotiate the Framework Convention, and kill the Kyoto Protocol.[4] The European Union's environment commissioner, Stavros Dimas, tried to contend that Kyoto would live on because its architecture would influence the shape of a new agreement, but developing countries, led by Egypt, were having none of it. They were not talking about preserving Kyoto's "architecture." They were talking about preserving Kyoto itself, in a second commitment period.

Bo convened another meeting on the Danish idea with key developed countries on November 20, and if you believed in omens, this was your day, because the tiny elevator carrying Sue, Jonathan, me, and my special assistant Clare Sierawski up to the prime minister's office got stuck. So much for American power. Sue's mood actually improved a bit despite her tendency to claustrophobia because she at least had something to worry about other than the resistance of key countries to do what we needed, though fortunately the shelf life of that improvement was not tested too far, and we emerged twenty minutes later to the amused smiles of our negotiating colleagues.

* * *

On November 25, President Obama announced a provisional US target of reducing greenhouse gas emissions "in the range of 17 percent below 2005 levels" by 2020, in conformity with US legislation "in the context of an overall deal in Copenhagen that includes robust mitigation contributions from China and the other emerging economies." Despite the carping we had heard earlier in the year, the US target was ambitious, comparable to the EU target if measured against a 2005 baseline. The Chinese answered the very next day with their own target to cut the CO_2 intensity of their economy—the amount of CO_2 produced per unit of gross national product—40 to 45 percent by 2020 compared to 2005 levels. We considered this a decent target, though not very ambitious.

* * *

On November 29, two days before the start of the last of Bo's meetings in the weeks before COP 15, he emailed me a story from the *Hindu Times*, dateline Beijing, reporting that China, India, Brazil, and South Africa—positioning themselves as a new group called BASIC—had agreed on their own ten-page draft declaration. Unnamed officials in Beijing explained that "the idea was to ensure that the western draft, which Denmark would unveil on December 1, 'does not get traction' to become the basis for the negotiations." The story noted that the idea of this "counter draft" came from China, and that it had prepared the initial text, with some revisions from the other three countries. Ramesh, ever dancing on a tightrope, made himself sound like a hard-liner, detailing the group's "non-negotiable demands" and declaring that he thought the new draft represented a "good starting point."[5] The BASIC group has been a significant, though often loose, negotiating alliance ever since.[6]

The December 1–2 meeting, once again in the former warehouse on the harbor, included roughly thirty countries, developed and developing.[7] It started with a bang. Minister Xie proposed that the draft China had prepared be used as the basis for discussion in the meeting rather than the Danish draft, but the draft included a full-on perpetuation of the firewall, and Bo quickly put it aside, turning to the new Danish draft he laid on the table. But the dissonant tone continued. Shinsuke Sugiyama, an always immaculately dressed and vocal Japanese diplomat who never tired of reminding his listeners that he was trained as an international lawyer, promptly declared that Japan had no intention of submitting a target for a second round of the Kyoto Protocol since that would amount to acquiescing in a continued firewall. Brazil's Luiz Figueiredo answered, "Should we go home now?" No one went home, but the two-day meeting made little progress, with no one giving ground on the contested issues. I spoke a number of times and tore into China for its untenable transparency position, which boiled down to self-reporting—"trust but don't verify," as Ronald Reagan might have summed it up. I bluntly told the group that countries had to decide what they wanted. The contours of a deal were clear enough: significant mitigation commitments by developed countries, significant mitigation actions pledged by major developing countries and submitted to the UNFCCC, genuine transparency, not limited to self-reporting, and financial and technology assistance. Countries would have to decide. Time was running out.

The positive takeaway from the meeting was that a sizable group of developed and developing countries had at least sat around a table for two days discussing and debating the elements of a proposed Danish political agreement. Bo read that as a hopeful sign of willingness to engage with the Danish idea at the COP, but Bo was overreading. That line from the *Hindu Times* article about making sure a "western" draft "does not get traction" should have given pause. The day after Bo's meeting, Xie and I met in Copenhagen, but made no progress on the two key issues dividing us—countries submitting their targets to the UNFCCC and subjecting their transparency reports to expert review. He again confidently assured me that we would work out the point about submitting targets to the UNFCCC ("internationalizing" them), and I again mistakenly read this to mean he would propose acceptable language in the end. But that never happened. In reality, he thought he could deflect my focus on this point, and I thought he was signaling flexibility, but we were both wrong.

* * *

As we made final preparations for the Copenhagen conference, we had to deal with an unanticipated curve ball that Norway accidentally threw when it announced on October 9 that President Obama was going to receive the

Nobel Peace Prize. The ceremony in Oslo was scheduled for December 10, while the leader segment of the COP was set for December 17–18. The White House, however, considered it out of the question that the president would travel twice within a week to the same small corner of the world, especially when the president's major health care legislation was still up in the air. Yet the Danes had bet the ranch on leaders attending the last two days of the Copenhagen conference. So we had a problem. I worked for several days with Mike Froman on our side and Bo on the Danish side to try to structure a meaningful visit the president could make on the front end of the conference before his stop in Oslo. But at a working dinner at Bo's December 1 preparatory meeting in Copenhagen, as I talked with a number of other delegates about the challenges facing the negotiations, it became starkly clear to me that it would be a serious mistake if President Obama did not come for the leaders' segment during the last two days of the conference. To give the UNFCCC its best shot at succeeding, we needed Obama in Copenhagen at the end of the conference as Denmark had intended. Coming to the COP when it mattered, even if it failed, would be far better than coming prematurely and being blamed for failure. I stepped out of the cafeteria where everyone was eating, and wandering in a deserted corridor, called my old friend Tom Donilon, Obama's deputy national security adviser. I told him why I thought we had to get the decision for an early visit reversed. President Obama had to come to Copenhagen when the fate of the conference might still be up in the air. This was a tough ask of Tom. The issue had been discussed and decided all the way up to hard-nosed chief of staff Rahm Emanuel. Tom nonetheless got the point, agreed to try, and delivered. The president's visit to Copenhagen was rescheduled for the end of the conference and that made all the difference.

3 Creative Destruction: The Battle

As COP 15 began its two-week run on Monday, December 7, its prospects were dubious at best. Negotiations under the formal track were going nowhere. The Danish idea of a short, legally nonbinding, prompt-start agreement was still controversial. And even if the Danish concept were accepted, there were still important issues up in the air such as transparency, the nature of differentiation, climate finance, and others. Then on Tuesday, the second day of the COP, the roiling disagreement over the Danish proposal burst into the open. The *Guardian*, a London newspaper that closely tracks climate change, reported in breathless tones that a "secret" Danish draft text had leaked. The article said that "developing countries reacted furiously to leaked documents that show world leaders will next week be asked to sign an agreement that hands more power to rich countries and sidelines the UN's role in all future climate change negotiations." A "senior diplomat" castigated the draft text as being "superimposed" without discussion in the normal UN negotiating process and claimed that it amounted to "a fundamental reworking of the UN balance of obligations" against the interests of developing countries.[1]

The *Guardian* story was grossly inaccurate and overwrought, but it succeeded in throwing the conference into disarray. The Danish idea of a short political agreement was no secret. The Danes had been talking about it for a couple months, Prime Minister Rasmussen himself spoke about it at length at the pre-COP on November 17, and a group of some thirty developed and developing countries had debated the particulars of a draft Danish text just a week before at the December 1–2 meeting that Bo chaired. But the leak was a well-timed missile that hit its intended target, blowing up Denmark's credibility and blocking its plan to shift the focus of negotiations toward its short-form proposal. Laurence Tubiana, the seasoned French diplomat who went on to become a leader of the 2015 Paris negotiations, pronounced "Bo's text" dead because it seemed to come out of the blue, failed to build on the negotiators' efforts over the past two years, and would get no traction with any of the key players Denmark needed to cut a deal.

I arrived in Copenhagen on Wednesday, December 9, in the midst of this turmoil, and soon after, held a press conference that wound up aggravating the already edgy tone between the United States and China.[2] After I had made it clear what the United States was prepared to do to get a deal along with what I thought would be needed from China and its emerging market allies, a reporter asked me whether the United States would provide financial assistance to China for climate mitigation. I said I thought public funds, which are inevitably limited, would be better spent on very poor countries, and that since China was a dynamic economy sitting on $2 trillion of foreign exchange reserves, it would not be first on our list for assistance. The Chinese were not happy. He Yafei, the smooth, senior Foreign Ministry diplomat on the ground in Copenhagen, was quoted by Reuters: "I don't want to say the gentleman is ignorant," he said, "but I think he lacked common sense or he's extremely irresponsible."[3] I wasn't concerned about He's comments. He was a skilled, accomplished diplomat whom I liked and respected. I took his reaction as a sign of just how sensitive China was about protecting its developing country status. Yet China's stunning economic rise and corresponding huge carbon footprint—US greenhouse gas emissions were around 1.4 times larger than China's in 1992, but China's emissions were more than 1.5 times larger than those of the United States in 2009—made it increasingly dubious for it to hide behind that shield.[4]

At any rate, between substantive disagreements on the tough issues and developing country anger at the Danes over the leak, the conference largely stalled out during the first week. There were plenty of meetings and consultations, but little progress. Starting on the weekend of December 12–13, the Danes tried one maneuver after another to get things going, but were repeatedly stymied by hard-line developing countries that wanted to prevent the Danish short-form idea from getting any oxygen. On Tuesday night, December 15, the G77 led by Sudan's Lumumba refused a private meeting proposed by Denmark's Environment Minister Hedegaard, then still the COP president. This was such a breach of normal COP etiquette that Bo worried a complete breakdown in the negotiations might be coming soon.

Of course, developing countries were not monolithic in their views. China was the most influential among them, with many followers in the G77, and China was quite prepared to let the negotiations fail with no real outcome beyond a promise to continue negotiating in 2010. But other developing countries like the Alliance of Small Island States (AOSIS), progressive Latin countries such as Chile, Colombia, Peru and Costa Rica, and a fair number from Africa and the least developed countries group (LDCs) were more interested in a deal in Copenhagen that would ensure real action to contain climate change and significant financial assistance, especially to help them manage dangerous climate impacts.[5] They favored a continuation of Kyoto, but were not as wedded as China and its allies to protecting the firewall.

The Chinese appeared to have two imperatives in mind. They wanted to prevent change to the firewall structure, avoiding unwelcome new provisions such as requiring an international review of their mitigation performance. But they also wanted to appear constructive and avoid blame if the conference failed. Of course, these two objectives were in conflict with each other since the first imperative could easily lead to a failed conference with fingers pointing at China. China's strategy appeared to be to work with allies in running out the clock so that the only available outcome would be a simple declaration that the parties would continue negotiating next year on the basis of the existing texts, even though those negotiations were visibly stuck. When Xie met with Connie Hedegaard, Bo, Sue, and me on Tuesday, December 15, he wouldn't budge. The next day, after having spent the conference working with allies to throw sand in the gears of every Danish effort to unstick the negotiations, He Yafei, in a demonstration of Chinese chutzpah, told Mike Froman that since there was no time left to negotiate the kind of political agreement that Denmark wanted, the only option was to issue a simple declaration forwarding the two working group texts ahead for further consideration in 2010.

That same day, December 16, having heard reports of a bogged down conference, Tom Donilon, the deputy national security adviser, called me. He wasn't happy. "Todd, what the hell is going on there?" he asked. "We don't send the president of the United States into meetings where we have no idea what's going to happen." Which was absolutely right and the way things were supposed to work. But the image that flashed through my mind as Tom talked was of a ski jumper who had already pushed off. There was no going back. All we could do was try to land the jump. I felt sure that the damage to the president would be worse if the COP failed and he did not come than if it failed after he did his best to save it. I told Tom what I thought, and he unhappily agreed. In his position, though, with an office just a couple dozen paces from the president's, he had every reason to be concerned.

At around midnight on Wednesday, with only two days from the scheduled conclusion of the conference, Mike, Sue, and I visited Prime Minister Rasmussen and his team in his Bella Center office. The room had the feeling of a wake. The three of us sat on a large couch across from Rasmussen, who seemed dejected and dispirited. The pressure on him was intense. He had taken the unprecedented gamble of inviting leaders to the COP for the first time on the bet that he'd have an agreement for them to sign, but now it looked as though he would have nothing. Some leaders had already arrived, more than a hundred would be there by the next day, and the negotiations were going nowhere. The prime minister's team looked worn out and deflated. Only Bo was still restlessly pacing around, consumed with the mission of finding a way out of their bind and getting a deal done. He held fast to two convictions, despite what he'd heard from China and others: first, that Denmark's only chance was to assemble a "Friends of the Chair" group, and second, that its objective still had to be some version of the Copenhagen Accord the team had been working on since September.[6] Bo was right on both counts. Countries may worship at the altar of a "party-driven process" and celebrate the transparency of negotiating sessions in large rooms with scores of participants, but in the end, tough, multilateral deals depend on high-level discussions in small rooms. And some version of the Copenhagen Accord was the only game in town since the conventional negotiating track was going nowhere.[7]

* * *

It was into this stressed environment that Hillary Clinton arrived early on Thursday morning, December 17. We met her in a conference room at our hotel, far from the lovely center of Copenhagen, in a not-yet-developed, gray, and desolate zone near the Bella Center. But she was full of energy, eager to hear our briefing, ready to go. And she came armed with big news. She was about to announce in a press conference that the United States was throwing its support behind the proposal for developed countries to mobilize $100 billion a year in financial assistance for developing countries by 2020, as originally proposed by UK prime minister Brown and more recently championed by Ethiopian prime minister Meles.[8] The United States had been wrestling with the finance question for some time, trying to avoid a sticker shock number for 2020 that might reverberate badly at home and complicate passage of our cap-and-trade legislation. Nevertheless, by early in the second week, Mike and I started talking about the need to embrace the Brown/Meles proposal, provided it was properly conditioned. We had already accepted the idea of a "fast-start" financing proposal for developed countries to provide public funding of approximately $10 billion per year from 2010 through 2012, but that wasn't enough to shift the dynamic of the conference.[9] Mike and Joe Aldy, a young economist and key member of the White House climate team, phoned home early in the second week and secured approval for the $100 billion pledge. We hoped that the secretary's news would jolt the COP and give developing countries a larger stake in getting a workable deal done. And it appeared to do just that.

At her press conference, Secretary Clinton explained the conditionality attached to US support for the $100 billion. The funds would come from a wide variety of sources, public and private, bilateral and multilateral, "in the context of a strong accord in which all major economies stand behind meaningful mitigation actions and provide full transparency as to their implementation."[10] She ended her prepared remarks quoting a Chinese proverb: "When you are in a common boat, you have to cross the river peacefully together"—a not-very-veiled appeal to China to recognize that the only safe path out of Copenhagen was for the United States and China to find common ground. Yet when I accompanied Clinton to her scheduled meeting with Premier Wen Jiabao at his hotel that afternoon, his aides told us he was "unavailable"—not exactly ready to cross that river together. The secretary had a series of other bilateral meetings during the day, a few of which I joined, including with leaders from the small islands and with Prime Minister Meles. Those two meetings in particular suggested that the frozen negotiation might be starting to thaw now that the largest-ever financial package for developing countries was genuinely on the table, with US support. Meles, a powerful voice, saw the chance of attaining his top objective.

The Friends of the Chair

Late Thursday night, Denmark finally managed to pull together a Friends of the Chair meeting, this time at the heads of government level, and that changed the game. The meeting started at around 11:30 p.m. in the Arne Jacobsen conference room on the second floor of the Bella Center, shortly after a dinner for heads hosted by Queen Margrethe. It was an extraordinary tableau, with world leaders gathered around a table not to ratify a deal that their ministers and negotiators had reached, or read prepared talking points, but instead to try to come up with some kind of agreement for a conference staring at outright failure without so much as a draft in their hands and only twenty-four hours left. Some twenty-five countries attended, including all the major players and representatives of most every country group. The lineup included UK prime minister Gordon Brown, German chancellor Angela Merkel, French president Nicolas Sarkozy, Ethiopian prime minister Meles Zenawi, Brazilian president Luis Inácio Lula da Silva, Australian prime minister Kevin Rudd, South African president Jacob Zuma, Mexican president Felipe Calderón, Maldives president Mohammed Nasheed, and Grenadian prime minister Tillman Thomas. Clinton represented the United States, since President Obama was not arriving until the next morning, and China, India, Algeria, Sudan (for the G77), Saudi Arabia, Colombia, and Lesotho were all represented below the leader level. Denmark's Prime Minister Rasmussen and UN secretary-general Ban Ki-moon sat at the front of the room. I sat behind Secretary Clinton, with Mike and Sue close by.

As the meeting began, it was clear that this gathering was the last chance for the Copenhagen conference to produce anything consequential. The Danes wanted to salvage some version of their Copenhagen Accord, as did we and most (though not all) countries in the room. Rasmussen presided, assisted by Bo. Brown spoke early and called for language embracing a 2°C limit on the rise in global temperature, a stated year by which global emissions had to peak, developed country emission targets and developing country emission actions submitted to the UNFCCC, transparency in the implementation of those targets and actions, continuation of the Kyoto Protocol, a pledge of $100 billion in financial assistance by 2020, fast-start financing of $10 billion per year from 2010 through 2012, and the preparation of a fuller, legally binding instrument in 2010. He urged that Denmark promptly bring together a group to prepare such a document.

Australia's Rudd joined in, calling for ambitious mitigation by all countries, with transparency, including verification, a strong financial package, and the continuation of the Kyoto Protocol. Merkel spoke along similar lines, urging the ministers and senior negotiators in the room—sometimes referred to as sherpas, top aides who support their leaders—to prepare a draft. Meles said that a political agreement of the kind Denmark proposed made sense to Africans as an approach leading to immediate action while also continuing negotiations for a legal treaty to be concluded at the earliest time possible. He stressed that the world cannot allow the rise in temperature to exceed 2°C and the $100 billion financial pledge had to be real.

He Yafei, representing China, complained that Premier Wen was upset because he had not been invited to this Friends of the Chair meeting. This was a canard because everyone was notified at the same time, during the queen's dinner, and Denmark never would have omitted China, whose agreement was crucial for any deal. Beyond this complaint, He repeated China's position that since there was no time left to negotiate a self-standing agreement, it would support only a simple declaration forwarding the working group texts for continued negotiation in the following year. Saudi Arabia, as long as it was led by Al Sabban, was happy to temporize, delay, and avoid real action. Brazil was ready to follow China in opposing the Danish plan, but was not as firm in its resistance. And Ramesh, for India, did not say much at this juncture, but earlier had amply demonstrated his openness to a Danish-style agreement.

President Nasheed—who had famously held an underwater cabinet meeting in October to dramatize the threat that rising seas pose to low-lying islands such as the Maldives—spoke passionately and pragmatically in favor of reaching a prompt-start, stand-alone agreement to begin making real progress quickly. And President Sarkozy, diminutive but on fire, stole the show by throwing a magnificent fit, pounding the table and nearly shouting that it would be completely crazy—"completment fou!"—for leaders to walk away without some kind of political deal. They would be ridiculed. Why were they invited here in the first place if there wasn't going to be something to accomplish? Well, there would have to be. He was furious, indignant, great.

At around 2:30 a.m., Rasmussen sent the leaders back to their hotels for a few hours of sleep, asking them to reassemble at 8:00 a.m. Sarkozy, exhausted and fed up, standing outside the conference center in the dark cold waiting for his car, turned to Clinton and said, "I . . . want . . . to . . . die." But the Danes went straight to work preparing a short text meant to capture the views expressed in this first session of the Friends of the Chair meeting, after which the sherpas would work on the text during the night. Bo reconvened the sherpas at around 4:00 a.m. with a draft text on the table, much of which the Danes doubtless had in their back pocket before the Friends of the Chair meeting started. The draft was short, only two and a half pages, organized into twelve principal paragraphs. This meeting did not make much progress, but at least allowed the sherpas to absorb what was in the draft and figure out what they thought needed further work. The discussion lasted until roughly 6:30 a.m. At that point, I returned to our delegation suite and slept on the floor of a small side room using my topcoat as a blanket before my team woke me up a half hour later to return to the action.

The leaders reconvened at around 8:30 a.m., finally getting an actual draft to look at. For the United States, the fundamental problem was Paragraph 5 on developing country mitigation and transparency. It said nothing about developing countries submitting or listing their targets or actions. And the transparency language on some form of international review was laughably weak—only that a party "may," "at its discretion," "answer a question" regarding information that it had provided about its actions. The Europeans as well as the island states and other developing countries that favored high ambition were also appropriately chagrined about the absence of vigorous collective targets for reducing emissions. They hoped to see a global commitment to reduce emissions 50 percent below 1990 levels by 2050 together with a developed country commitment to reduce emissions 80 percent by 2050, but neither was in the text.

* * *

At around 9:30 a.m., President Obama arrived at the Bella Center, and his presence set in motion a twelve-hour chain of events that proved pivotal. Mike and I briefed him in an odd, lower-level holding room that looked like it might once have been a dress shop, complete with mannequins. Standing around a high counter, we told the president where things stood generally and underscored our so-far-unsuccessful efforts to get the Chinese to accept adequate transparency language. We gave him an example of the kind of language we thought we needed, and to my surprise he said he wanted it still stronger. He told us that he was prepared to leave Copenhagen with no deal if it came to that, but was not prepared to take home language that was too soft. This was good news.

The president then went up to the Arne Jacobsen room with Mike and I trailing. When he walked in, the reaction was electric. Here was the smart, young, dynamic, groundbreaking US president who had accepted the Nobel Peace Prize just a week earlier. He had charisma and star power, and was charged up by the challenge of trying to land a deal on this last day in Copenhagen, with everything hanging in the balance. He took up his position in the US seat, with Clinton right behind him. I called his attention to the specific problematic language on transparency in Paragraph 5. After listening to the discussion for around fifteen minutes, he raised the US placard and took the floor completely on his brief. He started off broadly, then zeroed in on the importance of a mechanism to review whether countries were keeping their mitigation commitments, noting that without that kind of accountability, an international agreement wouldn't amount to much. He then eviscerated the existing text. Brazil's Lula repeated the point that China and India had been pressing for months—that there should be verification only of developing country actions that were financed by developed countries. Obama answered, "That would make sense if this was all just a complicated means of transferring money, of foreign aid, then Lula's framework would work. But I thought we were trying to fix the problem of climate change."

A while later, during a pause in the action, Obama got up and started working the room. He went and put his arm around Lula and chatted with him, then moved to a seat on the back bench next to negotiators as a small group clustered around him, and then continued on, working the room. About an hour later, he left to take his turn giving a speech in the plenary hall, after which he met with China's Wen. In preparing for the meeting, Obama wanted to discuss the specific language we needed for the review element of transparency, and I suggested that we bring Sue into the discussion. This led to Sue's out-of-body experience. She was upstairs working on text for Paragraph 5 of the accord with South Africa's Alf Wills and Australia's climate change minister, Penny Wong. Obama had asked his aide Reggie Love to go find her. Reggie was the all-purpose, multitalented, close-in aide who most presidents have referred to around any White House as the president's "body guy" because they're always with him. A twenty-eight-year-old, handsome, head-shaved, six-foot-four-inch double varsity athlete at Duke (basketball and football), Reggie was by a mile the coolest staffer in the Obama White House. Sue was poring over text with Alf and Penny when Reggie, who didn't know Sue, walked into the room and said, "Is Sue here?" She stopped poring, they locked eyes, and Reggie said simply, "Sue, the president needs you." At which point Sue more or less levitated out of the room to join our prep session.

At Obama's bilat with Premier Wen, Wen made a small concession. He said China would be willing to say that there should be "dialogue and exchange" at the international level regarding the transparency report that a developing country submitted. This was the first time China had opened the door to anything beyond self-reporting for "unsupported" actions. It fell well short of an international review, so Obama didn't accept it, but he encouraged negotiators from the two sides to keep talking, which Mike and He Yafei did.

In the midafternoon, Obama returned to the Arne Jacobsen room, which had grown heavy with a sense of frustration. Leaders had already been annoyed that He Yafai was representing China rather than Wen, but now a still less senior foreign ministry official represented China, Yu Qingtai. That Yu was gruff and irascible, with none of He's diplomatic finesse, didn't help matters, but the core problem was that he couldn't make decisions for China and repeatedly had to leave the room to call for instructions, never coming back with good news. The leaders in the room, including Obama, were sick and tired of this Chinese game. After an hour or so, Obama left, asking his White House team to track down Wen because there was clearly no way to get a deal done without talking to Wen directly. Hours went by with Wen nowhere to be found. Finally, in the early evening, when Obama was back in the Arne Jacobsen room for a third time, the White House team learned where Wen was, and Obama decided to go find him, invited or not. Before leaving, Obama gathered in a brief huddle with European leaders, including Merkel and Brown, who asked him to seek Wen's agreement to language calling for a legally binding treaty to be concluded by next year's COP. Obama said he'd try, though there was little chance of success, and then he took off, with Secretary Clinton by his side, and Mike and I following in a pack that included White House aides Denis McDonough, Robert Gibbs, and Jeff Bader.

The Crashed Meeting

It turned out Wen had company. He had convened a hastily organized meeting of BASIC countries that began between 5:30 and 6:00 p.m. India's Prime Minister Singh and Presidents Lula of Brazil and Zuma of South Africa were there, together with their top advisers. At the outset of that meeting, before we arrived, Wen indicated that the COP might fail and was concerned that some would try to blame that failure on BASIC. So the focus of the meeting, before Obama showed up, was not to work with the Americans or anyone else to complete a Copenhagen Accord but rather to prepare a statement to insulate BASIC from blame for the COP's failure.

A throng of reporters, TV camerapeople, and security staff were waiting outside the conference room where the BASIC meeting was going on. When they caught sight of Obama and his entourage moving rapidly toward them, something just short of all hell broke loose. The press was there to dutifully cover the BASIC meeting, yet all of a sudden—unannounced and uninvited—here came the US president with the secretary of state and a scrum of senior advisers heading straight for the door. TV cameras swung around, reporters scrambled this way and that, security tried to block the door to the room, and the only thing I had in mind was to get inside without getting clobbered by a camera.

President Obama, ever calm, collected, and confident, brushed passed security, opened the door, looked at the astonished faces of the four leaders and their aides, and said pleasantly, "Are you ready for me now or should we wait?" Premier Wen, having no alternative, invited Obama and his team in, people scrambled to reposition around the long conference table so the president and secretary could find seats, President Lula joked that "Barack" could sit with the Brazilian delegation, He Yafei graciously offered me an open seat on their side of the table, which I sensibly declined, and then it was game on. Obama sat directly across from Wen, with Lula and Zuma to Obama's immediate right and Secretary Clinton to his left. Prime Minister Singh sat just to Wen's left, while my counterpart, Minister Xie, sat three seats from Wen to his right.

After brief preliminaries, Obama said he wanted to talk about transparency, in particular the need for some kind of international scrutiny of the reports developing countries would submit on the implementation of their mitigation targets. After some initial discussion at the table, Mike and I,

India's Ramesh, He Yafei, President Lula's chief of staff, Dilma Rousseff, and Alf Wills from South Africa gathered in an extended huddle in a corner of the room, trying to find language that could work for everybody. Ramesh suggested "international discussions" and then "international consultations," but Mike and I vetoed both of those on the ground that the phrase needed to convey the idea of an assessment of some kind. Someone proposed adding the words "technical analysis," but I didn't like that because "technical" sounded like the exercise of bean counters rather than a higher-level assessment of how a country was doing on its pledges. Eventually, with no resolution, we moved the short distance back to the main table to brief our bosses on the state of play.

Ramesh took the floor for India and said that words such as scrutiny, review, or assessment were unacceptable to BASIC, and the best they could offer was "international consultations and technical analysis," but Obama was no more enamored of the word "technical" than Mike and I were. Suddenly, Xie, looking aggrieved, interjected that Senator Kerry, who had come to the COP, told him "dialogue and exchange"—the phrase Obama turned down in his bilat with Wen—would be fine. Obama answered, "Senator Kerry is not the president of the United States." Enough said. Ramesh then consulted with Wen and Singh, recommending that the BASIC side accept "international consultations and analysis," without "technical," provided Obama would agree to a reference about respecting national sovereignty. Wen proposed this, Obama had no objection, and so the transparency battle that had gone on all year was finally settled.

Mike and I then conferred briefly, and I told Secretary Clinton that we had one more important item to nail down, confirming that China understood it would have to submit its proposed targets to the UNFCCC secretariat, not just announce them in a white paper or press release—the issue of "internationalizing" that China had resisted all fall. With little hesitation, Wen said fine, at which point my friend Minister Xie blew his stack. He had been visibly steaming during the transparency discussion, especially when Premier Wen struck the deal. But once Wen took the further step of agreeing to submit China's pledge to the authority of the UNFCCC, Xie had all he could take. Red in the face, jabbing the air, speaking loudly and quickly in Chinese, gesturing alternately at Wen and Obama, Xie stopped the room. Obama, always the cool hand, looked calmly at Xie, waiting for him to finish. Wen, a pretty steady customer in his own right, looked a bit more startled, but not flustered. When Xie finished, Obama looked at Wen and asked mildly, "What's he saying?" Wen answered, "Not important."

What was important, however, and indeed crucial to the fate of the Copenhagen Accord, was that Wen made the strategic, improvised decision to compromise on matters of real significance to China. China had been a challenging counterpart during the entire COP and in fact the entire year, however well Xie and I got along, but at the moment of truth, Wen chose to be a leader and take the deal, even though it required acceptance of provisions that China had clearly intended to avoid and even though improvisation was very much not China's style.

Obama then raised the Europeans' point about setting next year's COP as the target date for reaching a legally binding agreement, but as anticipated, the BASIC countries flatly said no. They were not prepared to take the risk that a legally binding agreement would bind *them*. With that, the meeting ended. It had lasted about an hour and a quarter, and it made the Copenhagen Accord possible. Obama next met with his European counterparts to debrief them, after which they all returned to the Arne Jacobsen room at around 9:40 p.m. for the final session of the Friends of the Chair meeting.

* * *

Obama reported to the full group that the United States and the BASIC countries had reached a deal on the issues of transparency and the submission of mitigation targets. For the first time it looked as though Copenhagen would produce a real outcome—not what people had hoped for, but at least a real start. And then Obama left, advised by the Air Force One flight crew that they had to leave right away to get back to Washington before a huge snowstorm, later dubbed the Snowpocalypse, closed the city to air traffic. But there were still two issues to be resolved. The Europeans pushed aggressively to include a collective developed country target to reduce emissions 80 percent below 1990 levels by 2050, having abandoned their hope for a global reduction of 50 percent by 2050 in the face of Chinese opposition. To their surprise, however, China also blocked language for an 80 percent cut by developed countries. Chancellor Merkel was incredulous and angry, asking how China could possibly insist on deleting a sentence about what *developed* countries were committing to do, but the Chinese wouldn't budge. They had done the math and knew that an 80 percent cut by developed countries would still leave a heavy reduction burden for China if the world was going to meet the "below 2°C goal" temperature target that was also in the text of the draft Copenhagen Accord.

The last issue concerned the plea by the island states, led by the Maldives, to include language suggesting that the temperature goal might, in time, need to shift from "less than 2°C" to 1.5°C. The islands, especially low-lying ones on the leading edge of existential risk, believed less than 2°C was not good enough. Maldives president Nasheed argued passionately for 1.5°C language. China's Yu said no, but virtually everyone else in the room, developed and developing, was sympathetic to the islands' request, and eventually Yu gave way and language was crafted that everyone could support. It was circuitous, but at least captured the notion of a potential strengthening of the long-term goal, with an explicit reference to 1.5°C. And that effectively concluded the twenty-four-hour Friends of the Chair effort. But complications lurked around the corner in the plenary, where the accord still needed to be approved by the full COP.

The Plenary Revolt

We were well aware that the closing plenary was going to be a minefield given that decisions in the UNFCCC effectively require consensus. The COP had been in trouble from the start with the leak of the "secret" Danish draft and the extended refusal of developing countries to engage with Denmark. To further rub nerves raw, scores of negotiators from countries not included in the Friends of the Chair meeting were left to look after restless leaders who had journeyed to Copenhagen to sign a major new climate treaty only to find themselves lost in a conference on the verge of a nervous breakdown. As to the new accord itself, while it included elements sure to be attractive to many countries, it also swerved from the one true faith of the firewall, which the firewall's most devoted adherents would regard as blasphemy. And it didn't help that everyone was exhausted after twelve sleep-deprived days or that most country delegates had yet to see the accord.

Against that perilous backdrop, at 3:00 a.m., Prime Minister Rasmussen, as COP president, banged his gavel to start the proceedings, announcing that a Copenhagen Accord had been reached by a representative group of leaders. In a step tone deaf to the resentment coursing through the hall, he asked the parties to take an hour, read the text, consult among themselves, and then return to the plenary hall.[11] That's not what happened.

A handful of countries objected vociferously, led by Venezuela, Bolivia, Nicaragua, and Cuba from the ALBA group as well as Sudan, Tuvalu (the Pacific Island state), and Saudi Arabia.[12] An hour wasn't enough time. The public announcement of a deal before bringing the proposed text to the plenary was disrespectful (an angry reference to President Obama's briefing of the White House press corps before leaving). No one gave the Danish COP president a mandate to negotiate an agreement behind closed doors with a small number of countries and impose it on everyone else. The text itself was flawed. And so on. ALBA was doubly aggrieved for having been left out of the Friends of the Chair meeting, an inadvertent but costly mistake by Denmark and its UN advisors, though ALBA wouldn't have liked the accord in any event.

Emotions ran high right from the start, and Rasmussen's performance on the podium was telling. He was defensive in tone, unsure of himself whenever challenged by the rebels, and projected weakness. To get his attention, Venezuelan negotiator Claudia Salerno, with her well-known flair for the dramatic, waved a bloody hand—which she had cut herself—and said indignantly, "Do you think a sovereign country has to actually cut its hand and draw blood? This hand, which is bleeding now, wants to speak, and it has the same right of any of those which you call a representative group of leaders." Nicaragua intervened soon after and proposed that the accord be treated merely as a submission from the countries that negotiated it and then issued as a so-called miscellaneous (MISC) document, a UN bureaucratic designation that is equivalent to oblivion.[13] Yet Rasmussen, whether unaware of the consequences of the MISC status or not thinking clearly, proposed to do what Nicaragua suggested, though other countries objected to Nicaragua's proposal and he backed off. After Venezuela and Cuba lit into the accord and Cuba criticized Rasmussen himself, he said, "You're right, we can't by any means adopt this paper." It's true that Rasmussen couldn't ignore the quasi-consensus practice that governs COP meetings, but a strong president wielding the gavel leads the hall rather than getting led by it.

Sudan's Ambassador Lumumba, leader for the G77, took the floor with the most egregious intervention of the night. First, he disparaged the $100 billion assistance pledge in the accord—the pledge that was Africa's chief priority, and had been fought for and won by Prime Minister Meles—as a bribe. Then he shifted from insult to grotesquerie by calling the accord "an incineration pact" that "reflected the same set of values that tunneled six million people in Europe into furnaces." This, we may note, from a diplomat whose president, Omar al-Bashir, was facing an arrest warrant from the International Criminal Court for crimes against humanity in Darfur. Countries from the United Kingdom to Mexico, Canada, Grenada, and Norway called on him to withdraw his remarks, but he never did.

After watching Rasmussen's unsteadiness, I sent a message to Bo to meet me outside the plenary hall for a quick word. We huddled for a couple minutes, and I told him that I didn't think Rasmussen had the energy or grit to steer the accord to a safe landing. I didn't know whether he was exhausted, in over his head because he lacked international experience, or had just lost his fighting spirit, but he looked beaten down. I worried aloud that he wouldn't be able to manage this unruly body with the strength, finesse, determination, and command needed to get the job done. I suggested that at some point, it might make sense for him to take a little walk and temporarily hand the gavel off to someone else.

Back in the hall, I walked the few feet over to the UK delegation and told my friend Pete Betts, their lead negotiator, that we needed to get his minister, Ed Miliband, back in the hall right away. Ed was strong and effective, spoke with clarity and conviction, and was a key ally who would stand loud and strong for the accord. He rushed in a few minutes later, looking like a man who hadn't quite finished dressing, with a jacket thrown on, no tie, and hair disheveled, but fully ready for battle. In short order, he took the floor and made a powerful, dramatic intervention in the best tradition of British leaders. He said that the institution of the UNFCCC and the work we were all doing faced a profound crisis with two roads ahead. We could embrace the accord, a document produced in good faith, not perfect, but an important start that would get mitigation moving and climate funds flowing, or we could reject the accord and wreck the conference. Dessima Williams, the tough, intelligent, fearless negotiator from Grenada, speaking for AOSIS, took indignant exception to the idea that there was something inappropriate about the Friends of the Chair leaders striving to produce an outcome. She spoke directly to the opposition and said, "I can't listen to my prime minister and my long-tired self getting discredited. I find it offensive. And I call on my brother from Sudan to get hold of your emotions. We have to help each other, not condemn each other."

I took the floor, defended the representative process that Denmark followed to produce the accord, underscored President Obama's direct and successful engagement to bridge differences alongside the leaders of China, India, Brazil, and South Africa, and highlighted the major advances the accord made in terms of reducing emissions, providing transparency and accountability, pledging financial and technology assistance, and increasing the COP's focus on adaptation and deforestation. I said it would be disturbing to walk away from all of this, both for the planet and the ongoing health of the UNFCCC, and endorsed Miliband's call for adopting the accord.

Overall, the accord had the widespread support of the vast majority of countries, developed and developing. Ethiopia supported it on behalf of the African Union, and Algeria did the same on behalf of the African Group of Climate Change Negotiators. The LDCs supported it. AOSIS, the island group, supported it. Mexico and other progressive Latins supported it. Among the BASICs, Brazil and India both supported the accord, and South Africa effectively did since the African groups never would have spoken up if South Africa did not approve. Yet for all of that, we were not winning. The math was on the side of the protesters because of the UNFCCC's quasi-consensus practice. Moreover, the applause when protesters spoke made it clear that the group had many sympathizers in the room, among both countries and observers. The ALBA countries and Tuvalu kept pushing the Nicaraguan proposal to bury the accord in MISC status. Saudi Arabia jumped in to castigate the plenary as the worst ever and declare that since there was no consensus, Rasmussen had no choice but to find a procedural solution and move on. And China was conspicuously quiet, despite the pivotal role played by Premier Wen in making the accord possible.

President Nasheed of the Maldives, passionate, lucid, and pained, made a fervent plea to his fellow developing countries to keep the accord alive. Philip Weech of the Bahamas followed suit and chastised the rebels, saying that it was "totally unacceptable" for President Nasheed to have to "beg for the document to be adopted," and reminded delegates that MISC status meant that none of the elements of the accord would become operational, including in particular its financial pledges. Ed Miliband then reiterated an idea presented earlier by Slovenia that the accord be adopted while allowing countries that objected to be recognized in the formal adopting decision. Rasmussen, ever cautious and hesitant, said, "I want to be sure that no one will be against this"—a sentence that no chair should ever utter. Of course, the hard-liners were against it. Rasmussen said he really regretted that the accord couldn't be adopted, but then, trying one more fallback, suggested with no clarity that perhaps countries could indicate their support for it at some later time, yet before he even completed his sentence, the protesters started hooting and hollering and clacking their placards, at which point Rasmussen gave way again, saying, "So we can't adopt this paper and can't in any formalized way register which countries would subscribe to it."

At that moment, around 8:00 a.m. on Saturday morning, December 19, I thought Rasmussen had given up the ghost and was about to let the accord—Denmark's own hard-won accomplishment—sink beneath the waves, never to be heard from again, leaving the COP a dismal, outright failure. Knowing at that urgent moment that it would be better for the United Kingdom than the United States to speak, I grabbed Ed and said we needed an immediate pause in the proceedings to give us time to figure out a way forward before Rasmussen brought down his gavel to end the discussion. Ed requested the floor, was recognized, and said simply, "The United Kingdom asks for a short adjournment." Rasmussen granted it.

The next moments went by in a blur. I have a rare but explosive temper, and at that instant, when everything we'd worked for was threatened by what struck me as the unconscionable passivity of those on the podium, I lost it. Rasmussen, the support staff from the UNFCCC secretariat, and the secretary-general himself—who had commendably made climate change a high priority—seemed to me defeated, inert, and prepared to let the spoilers have their way. This was not leadership, and it broke faith with the agreement reached by the Friends of the Chair leaders. We had fought all year, the issue was critical, my president and secretary of state had gone all out, we had taken a critical step, and the idea of losing it all was intolerable. We simply had to fight back. So, following instinct and rage, working on two thirty-minute naps on a hard floor in the previous forty-eight hours, I took off down the broad aisle toward the podium only vaguely aware of the phalanx of others that started following me. When I got there, I exploded. "This is a disgrace! Shame on you!" I shouted, pointing at the secretary-general of the United Nations and those around him. Whether that galvanized action or only incredulity I don't know, though Bo told me days later that he thought the shock of it actually helped move things during the hours that followed.

At any rate, it became clear not long after my outburst that Prime Minister Rasmussen had in fact stepped away and was going to take a break. An active effort began to find a strong, respected replacement to run the meeting until Rasmussen returned. Someone suggested Christiana Figueres, a forceful, respected Costa Rican diplomat then serving as vice president of the COP Bureau (an administrative body) and later to become the executive secretary of the UNFCCC. Christiana seemed perfect, she said yes, and everyone was on board. But from the "you can't make this up" file, it turned out that her term of office in the bureau had expired eight hours earlier, at midnight, so she was no longer eligible to take the gavel. In short order, someone suggested, Philip Weech, the strong Bahamian lead negotiator. He said yes and proved more than equal to the task. Meanwhile, a clutch of negotiators had formed during the pause to try to work out exactly what should be done with the accord. Our laser focus was on giving it some kind of recognition and standing so that it didn't just fade away. Two of our star young State Department lawyers, Jeff Klein and Kevin Baumert, were in the middle of the scrum and either suggested or supported saying that the COP would "take note of" the accord, and that suggestion stuck. Meanwhile, Bo, fuming about China's conspicuous silence and never shy about expressing himself, found Su Wei, Minister Xie's deputy and the lead Chinese delegate in the room, and let him have it: "I hope you know what you're doing," Bo said, with his inimitable, ringing cadence. "In our system, I would get into trouble if I allowed a deal my prime minister had just personally agreed with the American president to crash. They wouldn't take that lightly. But maybe it's different in your system." Su didn't answer.

The plenary resumed around 10:30 a.m., with Weech in the president's chair. He ran the session like a commander—he knew what he had to do, he did it, and he didn't ask permission. He told the hall that the proposed decision on the table was that the COP "take note of" the Copenhagen Accord. With no debate and hearing no objection, in a matter of seconds, Weech banged his gavel, and the deal was done. Representatives of countries all around the plenary who had fought for strong climate action, developed and developing, including an emotional President Nasheed, stood and applauded. I took the floor a bit later to urge the secretariat to notify countries of exactly how they would be able to "associate" themselves with the accord—something that Weech had referred to, but not, to my mind, with enough clarity. The accord hadn't been adopted, but I wanted the next best thing: to get as many countries as possible to declare their acceptance of it and submit their mitigation targets and actions. Weech agreed that the secretariat would provide guidance.

And so the struggle in Copenhagen ended, though the battle to keep the Copenhagen Accord alive had just begun. Weeks later I sent the secretary-general a personal letter of apology for my undiplomatic outburst at the podium. I was sorry to have yelled at a good man who was devoted to the climate cause and doing his best. But I didn't regret it.

Postmortem

Around the world, the Copenhagen conference was broadly dismissed as a flop. Andreas Carlgren, the Swedish environment minister while Sweden held the presidency of the European Council during the second half of 2009, described COP 15 as a "disaster" and "really great failure."[14] This disappointment, shared widely, if not so dramatically, among Europeans, was partly a product of outsize expectations, partly the frustration of seeing China and allies block even relatively modest expressions of ambition, partly the belief that nothing short of a legally binding agreement could be taken seriously, partly the aggravation of dealing with a United States that had its own, quite different approach to the negotiations, and partly annoyance at President Obama's parting words to the press, which they interpreted as taking credit for the last-minute deal that they felt should have been shared with them.[15]

Beyond Europe, among a broad swath of developing countries, including many that supported the Copenhagen Accord in the final plenary, the perception of Copenhagen was no more charitable. Muhammed Chowdhury from Bangladesh, a lead negotiator for the G77, said, "The hopes of millions of people from Fiji to Grenada, Bangladesh to Barbados, Sudan to Somalia have been buried. The summit failed to deliver beyond taking note of a watered-down Copenhagen accord reached by some 25 friends of the Danish chair, heads of states and governments. They dictated the terms at the peril of the common masses."[16] And this assessment of Copenhagen as a failed conference still persists in many quarters.

But the assessment was fundamentally flawed. The Copenhagen conference was bitter, turbulent, acrimonious, and teetered at times on the brink of collapse. Both President Obama and Secretary Clinton came away with a kind of headshaking wonder at what they had just experienced. Clinton called it both the most extraordinary thirty-six hours she had spent in public life and the most chaotic political event she had witnessed since seventh grade student council. But the US side never saw Copenhagen as a failure. We knew the accord was an important document, that it began an essential pivot away from the old firewall paradigm and was a potentially significant step forward toward an effective international response to climate change. Others saw progress in Copenhagen as well. The United Kingdom's Miliband expressed that view in a piece he wrote for the *Guardian* the day after the conference adjourned.[17] And a variety of US commentators appreciated that something important had just happened.[18]

For the first time, the COP endorsed the goal of holding the increase in the global average temperature below 2°C and agreed to consider strengthening that goal even to 1.5°C following an assessment in 2015. For the first time, many developing countries, including all of those in the MEF, agreed to specific targets and actions to limit their emissions, and agreed to submit those targets and actions to the secretariat.[19] For the first time, developing countries agreed to be subject to a transparency regime including not just self-reporting but also some form of international review of their emission inventories and actions to limit emissions, whether those actions were financed by developed countries or not.[20] And for the first time, the legal treatment of developed and developing countries was symmetrical—politically, not legally, binding for both groups.[21]

At the same time, the accord included important new elements for the benefit of developing countries. Developed countries pledged to provide "fast-start" funds of $30 billion in the three years 2010–2012 and mobilize $100 billion per year by 2020, conditioned on "meaningful mitigation actions and transparency on implementation" by developing countries. The accord also called for the establishment of a High-Level Panel to consider ways to meet the $100 billion goal, a new Green Climate Fund, and a new body to accelerate technology development and transfer. And it called for enhanced focus and action to build the capacity to adapt to climate impacts as well as prevent deforestation.

We didn't know where things were going next, but we knew that Copenhagen broke new ground. We were certainly disappointed that the accord had not been formally adopted, but it had survived, and the secretariat had agreed to establish a process for countries to formally "associate" themselves with it. However far the accord fell from the hopes and dreams of many, it had moved the ball significantly down the field.

The assessment of Denmark's performance in the aftermath of COP 15 and to this day has been routinely harsh. It was castigated for producing an accord that did not emerge directly from the formal negotiation groups, contrary to the UNFCCC's devotion to a "party-driven process"; for running a secretive, noninclusive process that left out developing countries; for being too close to the United States; for inviting leaders to meddle in the business of negotiators; and for losing control. These grievances are mostly about process, and there is no question that Denmark made process mistakes along the way, especially in the nature and timing of its consultations with developing countries in fall 2009, leading up to the COP. But it is also true that countries became angry at Denmark much more because of *what* the Danes were trying to do than *how* they did it. Countries were upset because the "politically binding" operational agreement Denmark started pushing in the fall departed from the prevailing orthodoxy of climate negotiations. Denmark could have played a perfect hand as far as process went and there still would have been turmoil and strife to get the Copenhagen Accord done. There was simply no way to start shifting a paradigm without breaking china.

Moreover, whatever mistakes of execution the Danes made, they understood two crucial things and had the nerve to act on their understandings. They understood that the old orthodoxy could not deliver anything in Copenhagen, so they developed their politically binding option to make an immediate difference in the fight against climate change while keeping the possibility of a legally binding agreement alive in the years ahead. And in pursuing this option, they sought to thread a needle between what the three dominant players—the United States, China, and Europe—could plausibly accept. This was a high-risk strategy that made the Danes widely unpopular, but they were willing to roll the dice. They also understood that landing a paradigm-shifting deal would take leaders; negotiators would be too stuck in their ways and too risk averse to do it. So the Danes took the audacious gamble of turning the COP into a leaders meeting, even though the prospects for success were anything but assured. In short, Denmark—above all Prime Minister Rasmussen and Bo Lidegaard—went for broke, nearly crashing the whole process, but in the end producing the Copenhagen Accord, the first crucial step on the road to Paris.

A number of other countries played important roles in getting the Copenhagen Accord done and keeping it alive. No country played a more varied and unusual part than India. In its traditional posture, India was devoted to the firewall, and this was what we saw in the first half of the year at our MEF meetings, embodied in Prime Minister Singh's adviser Shyam Saran. But India's undaunted, iconoclastic environment minister, Jairam Ramesh, had a different approach in mind. He managed to stay on the balance beam between being a member of BASIC, mindful of India's own climate orthodoxy, and nudging his country and its allies toward a new climate modus vivendi. He opened the door to an eventual agreement on transparency by the remarks he made at the Arrowwood meeting in Westchester, New York in September, by walking with me around the lakes in Copenhagen, by speaking out at the pre-COP meeting, and by playing a crucial role in the famous session between President Obama and the leaders of China, India, Brazil, and South Africa, where the transparency deal was finally struck.

Ethiopian prime minister Meles, the strongest voice for Africa in Copenhagen, made it clear at my meeting with him in November that he was comfortable with the Danish notion of a short, political agreement so long as certain conditions were met, and when he saw that they were, during the Friends of the Chair meeting in Copenhagen, he blessed the idea of the Copenhagen Accord and thus legitimized it for much of the Africa group. Small island leaders—led by the Maldives' impassioned President Nasheed, joined by such determined voices as Dessima Williams from Grenada and the Bahamas' Philip Weech—stood up and demanded support for the accord as an important, even if insufficient step forward. The Europeans, pained as they were by what they saw as the accord's inadequacies, made the right call to embrace it and fight for it in the end. Leaders, including the United Kingdom's Brown, Germany's Merkel, Sarkozy from France, and Rudd from Australia all fought for as ambitious an accord as possible. And of course, the United States' early, unwavering recognition that the paradigm we inherited was untenable for both political and substantive reasons, aided by Australia's schedules paper, helped prepare the way for Denmark's move toward the Copenhagen Accord.

Beyond the important content of the accord, the struggle over it had other lasting ramifications. Major developing countries such as China, India, and Brazil had always been vehemently opposed to subdivision of the overall developing country group. Remaining in the same indivisible category as the poorest countries protected the majors against obligations and expectations while also enhancing their coalition power. Yet in Copenhagen, the interests of the developing majors and developing vulnerables and progressives started to visibly diverge. The islands and their progressive allies pushed for maximum ambition while the majors were focused on protecting the firewall and minimizing expectations aimed at them. The G77 did not break apart, but there were new fissures in the alliance, and in the years to come, this led to new coalitions and new strategies.

The Copenhagen COP had a lasting impact on China's climate change diplomacy too. Copenhagen was a problematic conference for China. As noted, China entered the COP with conflicting objectives: don't let the Danish idea for a short, politically binding accord get done, but don't get blamed if the COP fails. And in the end it didn't achieve either objective. The Copenhagen Accord did get done, and even though not adopted, was alive to influence the future. And at the same time China was blamed not for the perceived failure of the COP but instead for having opposed the interests of poor and vulnerable states for which climate change is a mortal threat.[22] China also fought openly with the United States, to no good effect. Its president, Hu Jintao, declined to come to Copenhagen at all, and China angered a conference room full of national leaders by sending bureaucrats rather than Premier Wen to the crucial Friends of the Chair meeting. This conduct, while serving China poorly, flowed from deep-seated Chinese convictions. China's priority in UNFCCC negotiations had always revolved around preserving its own developing country status and the special, lenient treatment that entailed. This priority in turn linked to protecting China's capacity to manage its own economy as it sees fit, with no outside interference, which in turn linked to the Chinese Communist Party's core bargain with the Chinese people: its one-party license to rule premised on continually improving standards of living. And all of these linked to China's fierce resistance against any perceived encroachment on its sovereignty—an obsession tracing back to the humiliating era of "unequal treaties."

In view of China's conduct in the following years, I came to believe that the Chinese decided to modify their climate approach after Copenhagen in ways meant to avoid getting into public discord with the United States, or getting crosswise with poor and vulnerable countries, while still preserving their core priorities as best they could. While Minister Xie and I had already developed a good and cordial relationship in 2009, from 2010 forward he put an even greater emphasis on good climate relations between our two countries, on genuine effort to work out our differences, and on not criticizing each other publicly. We both always recognized that we had significant substantive disagreements, but we also knew that our two countries occupied a special place in climate diplomacy and would have to find ways to work together productively. Finally, China took care not to get out of sync with vulnerable countries again, though this became more challenging over time, as their substantive interests continued to diverge.

4 The Thirteenth Month

The tumultuous events at the 2009 Copenhagen COP, ending in the near death and partial resurrection of the Copenhagen Accord, were so dramatic, contentious, and consequential that the conference continued to exert its gravitational pull on the climate negotiations all through 2010. As we saw it, the Copenhagen Accord struck the first blow against an outdated paradigm and represented the fragile beginnings of a new mindset in international climate negotiations. But while "taken note of," the Copenhagen Accord was still not an official UNFCCC document, adopted by the parties. Thus our strategic objective for the year was to safeguard the progress that the accord represented by embedding its main elements in a formal COP decision adopted at the 2010 COP in Cancún. For any number of developing countries that distained the accord, including China and its allies, their principal objective was evidently to bury it, or at least those elements that started chipping away at the firewall. For Mexico, the incoming COP presidency, the Cancún COP essentially presented a challenge akin to the one faced by all the king's horses and all the king's men: to put the UNFCCC back together again.

Our own first step was to make sure that more than a hundred countries "associated" themselves with the accord and all the major developing countries submitted their mitigation targets to the UNFCCC. Denmark and the UNFCCC secretariat, among others, pushed in the same direction. In the first weeks of January, I spoke to representatives of eighty to ninety developing countries in New York and Washington about the importance of associating with the accord. My message was simple. However problematic COP 15 might have been as a conference, key elements of the Copenhagen Accord—including the goal of holding the global temperature increase below 2°C, mitigation, transparency, financial and technology assistance, a new Green Climate Fund, and an enhanced focus on adaptation and forestry—were vitally important to us all. If countries wanted to see them preserved at COP 16, they were going to have to stand up and be counted. My biggest concern initially was about China and its BASIC allies, but they all submitted their targets by January 31, as called for in the accord. While China, India, and Brazil declined to formally associate with the accord, they agreed to be listed on the first page of the official version, together with 111 other countries—the next best thing.[1]

Still, in seeking to incorporate the substance of the accord into the outcome of the COP in Mexico, we faced strong headwinds. Adoption of the accord had been blocked in the final plenary session in Copenhagen significantly because important countries opposed its initial pivot away from the firewall and it was far from clear that this opposition could be overcome in 2010. Such uncertainty led Ed Miliband to ask me in early February whether I was giving any thought to what might happen if the COP in Mexico came undone—in effect, what a plan B might look like. Our shared assumption was that if the COP stalemated in Mexico, we might well be forced to find a new approach to tackling climate change outside the UNFCCC. The United States never did work on a plan B because plan A, the business of trying to make COP 16 successful, consumed our energy. But concern about the vitality of the multilateral regime in 2010 was a very real a factor in the negotiations.

* * *

The backlash against the Copenhagen Accord was visible in the year's first intersessional meeting in April, when China and others attempted to block any consideration of the accord in discussions about a 2010 outcome at the Cancún COP. And a variation on that theme played out in the first of what became an annual meeting of around fifty ministers hosted by Germany, called the Petersberg Dialogue. There, Minister Xie's number two, Su Wei, came up with one of his patented reinterpretations of words, disputing the concession Premier Wen made in Copenhagen that the transparency clause negotiated in the famous final meeting with President Obama—international consultations and analysis—applied whether a developing country's action was supported by outside financial assistance or not. Su was his own legend in climate circles. Small, bespectacled, with hair that stood on end, and a slightly shuffling gait, he deployed perfect English with astonishing dexterity, was unfailingly soft-spoken and measured, and projected complete confidence, even when employing prestidigitation to get out of something China previously agreed to. When Su starts a sentence with "it is very clear," you know you need to be on your guard.

Yet while some developing countries were trying to pull back from the Copenhagen Accord, the new tensions visible in Copenhagen within the broad G77 led other developing countries down a different path. Negotiators from the United Kingdom (Pete Betts), Colombia (Andrea Guerrero), and Australia (Howard Bamsey) had been struck by the dysfunction of the talks in Copenhagen, where there seemed little room for constructive discussion. So in March, the three countries collaborated on a first meeting of negotiators from roughly thirty developed and developing countries in the Colombian city of Cartagena. This meeting proved to be a consequential strategic move, the first step in building an effective coalition among developed and developing countries committed to ambitious emission reductions and a legally binding treaty. Participants in the new "Cartagena Group" came from Europe, Latin America, Africa, Asia, and the island states.[2] The new group met three times each year, always in developing countries, and aimed to work out negotiation compromises and solutions. The United States was not invited to join partly because our posture on legally binding emission targets was out of sync with the group and partly because the Europeans were happy to make progress in a venue where they could operate outside the long shadow of the United States. What the United Kingdom, Australia, and Colombia understood was that progressive developing countries more interested in stopping climate change than in defending the firewall needed a place to pursue their agenda outside the rough environs of G77 coordination meetings, where the bigger or angrier countries often dominated. In short, Cartagena was the first concrete sign of the division among developing countries that began to manifest itself in Copenhagen.

* * *

Meanwhile, it was a stroke of good fortune in 2010 that Mexico held the COP presidency. Mexico was blessed that year by having a president, Felipe Calderón, who was a firm believer in the importance of containing climate change; a foreign secretary, Patricia Espinosa, who was similarly committed; and a leading diplomat, Luis Alfonso de Alba, with a deep reservoir of multilateral experience, a sharp tactical mind, and a steady disposition.[3]

From the earliest days of 2010, he went about trying to settle jangled nerves by ensuring that Mexico would run an open and inclusive process, and would not develop its own plan or text. He had good historic relationships with the ALBA countries, and made sure to include, listen to, and respect them. De Alba first came to see us in Washington in January as part of a broad diplomatic mission to hear what countries were thinking about COP 16 and reassure them that this year would be different. Immaculately dressed, with a trim beard, bald head, and whenever possible, cigarette in hand, he radiated a sense of calm, patience, wisdom, and reserve, and held his cards close. He seemed not to lean for or against us but rather to understand our concerns. I thought he was someone we could work with.

The other stroke of good fortune in 2010 was the appointment of Christiana Figueres of Costa Rica as executive secretary of the UNFCCC, replacing Yvo de Boer, who served from 2006 to 2010. Christiana, the one who had nearly taken the gavel during the last hard night in Copenhagen, brought long experience in climate negotiations to the job as well as her special brand of energetic, never-say-die optimism. She was her predecessor de Boer's antipode. De Boer, smart and deeply dedicated, is serious, dour, even a bit grave, and brought to the table his own views about how the negotiations should or should not proceed. Christiana, a five-foot dynamo with a quick laugh and ready smile—except when you don't please her—is a walking embodiment of the power of positive thinking. She's unsinkable, uninhibited, indefatigable, a motivator, and a get-it-done pragmatist who doesn't care so much whether we go this way or that as long as we get there. Legally binding? Fine. Not legally binding? Also fine if it gets the job done. She has one brown eye and one blue eye (actually), every variation on a theme is welcome in her house, and she is equally comfortable with a footman or a queen. She arrived at a good moment and stayed until she helped get a history-making agreement over the finish line in Paris.

* * *

In the summer, US stature took a hit when the Senate finally abandoned its effort to pass its own version of the Waxman-Markey cap-and-trade legislation after coming one vote shy of the sixty needed to break a Republican filibuster. It was a bitter pill for Democrats and climate activists at home, and hurt us globally since it suggested that the United States would not be able to meet its 2009 emission reduction target. Mindful of the diplomatic importance of *not* walking away from our target, Secretary Clinton, Deputy Secretary Jim Steinberg, and I recommended to the White House that we stick with it, even though it had been premised on Waxman-Markey. President Obama agreed, saying a few months later that cap and trade "was just one way of skinning the cat . . . not the only way."[4] But our failure to enact the comprehensive, binding cap-and-trade legislation, with ambitious emission markers all the way to 2050, left countries disappointed and wary. We used President Obama's talking point and tried to reassure countries that not much had changed, but the failure of the bill inevitably undermined our credibility.

Fall Preparatory Meetings

As summer turned to fall, the mood of the diplomatic climate community seemed uneasy given worries about the lack of progress to date and the consequences for the multilateral climate process if a second COP in a row seized up. At our MEF meeting in New York in September, Dan Reifsnyder, a well-traveled State Department climate veteran substituting for Mike Froman as chair, captured this unease in his opening remarks, saying, "The context of this meeting is a sense shared by many that the negotiations are not on a path to success." During an early September trip to Geneva with Jonathan Pershing for a climate finance meeting, Jonathan and I had dinner with France's lead negotiator, Brice Lalonde, at the French ambassador's residence. Lalond, echoing what Miliband asked me back in February, said that France was thinking about what a post-UNFCCC world would look like if the UNFCCC process failed and also reflecting on the use of trade measures. He added that other European countries were having the same kind of quiet conversations. The next morning, Jonathan and I had breakfast with India's Ramesh before the start of the climate finance meeting, and he, too, was concerned about whether the UNFCCC could produce meaningful results. On the way to Geneva, I had stopped in Brussels to have lunch with a wise EU friend from 2009, Karl Falkenberg, who summed up the underlying problem in simple terms: the United States would not accept the old, bifurcated paradigm, but China and its allies would not give it up. Solving that riddle, step by step, was a crucial part of our work for seven years.

The most important and contentious debate in 2010 involved what issues would be included in Cancún out of the basic bargain of Copenhagen: on the one side, stronger mitigation and transparency provisions applicable to at least the major developing countries, and on the other, significant provisions on financial and technology assistance as well as attention to adaptation and forestry. By September there was a growing chorus arguing that we should first harvest the low-hanging fruit—issues where there was already a good deal of agreement, such as financial and technology assistance, which developing countries cared most about. By this logic, the more contentious issues, such as mitigation and transparency, could be left for another day. We heard this line not only from firewall proponents but also from progressive friends like Barbados, Singapore, and Germany, which wanted to avoid having the benefits of financial and technology assistance caught up in a firefight over mitigation and transparency. I made it clear that the United States wanted no part of the low-hanging fruit approach. Our leverage to secure a good outcome in Cancún depended on keeping the important elements of the Copenhagen bargain yoked together in what came to be called a "balanced package." If we gave that leverage away in 2010, we would be hard-pressed to advance the paradigm shift that began with the accord.

* * *

From October 4 to 9 in Tianjin, the Chinese hosted the last intersessional meeting of the year. It was a problematic gathering. We and our allies expected the emission targets submitted after Copenhagen to be formally included in the Cancún outcome, together with other important issues in the accord. But contrary to what had been agreed to in Copenhagen and written into the Copenhagen Accord, Su Wei told Jonathan Pershing that China opposed any formal submission of mitigation targets as part of the Cancún outcome. He also told Jonathan that the "agreed structure" was for developed countries to take legally binding targets while developing countries reported on their nonbinding actions after the fact—a structure flatly rejected in Copenhagen. And China's transparency negotiator, Teng Fei, told Kate Larsen on our team that international consultations and analysis applies only to the limited question of whether a country followed technical reporting guidelines in preparing its "national communication," not to how the country was actually doing in implementing its targets or actions. This didn't pass the straight-faced test. We did not do battle on transparency for most of 2009, taking the matter all the way to

President Obama's final meeting with Premier Wen and the other BASIC leaders simply to secure a technical check that reporting guidelines had been followed.

I thought China's conduct in Tianjin was so problematic that I needed to call it out publicly for backtracking from what we all agreed to in Copenhagen, so I used a speech I was giving at the University of Michigan on October 8 to do just that. On October 19, I flew to Beijing for a two-day bilateral visit with Minister Xie. After a nice half day during which he took me on one of China's high-speed trains to visit his hometown of Tianjin, we returned to Beijing and got down to business. He brought up my Michigan speech right away, aggrieved by my public criticism, which is exactly what I wanted. In the whole first part of our conversation, he focused on getting me to admit that China had not backtracked, but I wouldn't agree. He said that the next time we hear "rumors," we should exchange views face-to-face since we're good friends and not criticize each other publicly—a refrain he repeated for more than a year whenever we met.

At the end of the first day's talks, Xie did one of his little summaries that started with "You don't think we're backing away" and ended with "I hope we find an appropriate solution" to the mitigation issue we had been discussing. "More tomorrow?" And I—perhaps tired, jet-lagged, or just momentarily lacking my usual vigilance, simply said "yes," by which I meant "more tomorrow." But afterward, Sue called to my attention that Xie might have taken my "yes" as agreement that I didn't think China was backing way. So I started off our session the next morning with a clarification: "You said we agreed that China was not backing away from the Copenhagen Accord. We don't disagree or agree," meaning it would depend on China's behavior. He replied, "Every time during our discussion I sum up." I interjected, "You are very good at summing up *your* interpretation." He smiled and said, "Every time I sum up you agree, then sleep and disagree." Which was not true but amusing.

We then engaged in a lengthy conversation in which Xie's main purpose was to get me to agree that there were three categories of countries in regard to mitigation: Kyoto Protocol developed countries, which make legally binding commitments; the United States, which under the 2007 Bali Action Plan, was supposed to show effort "comparable" to other developed countries; and developing countries, which had the choice to take "voluntary" actions. My main purpose was to get him to agree that for those developing countries that did pledge under the Copenhagen Accord to submit their targets or actions to the UNFCCC secretariat, the treatment of developing and developed countries was parallel, both agreeing to implement what they had pledged. After a lengthy back and forth, we each more or less agreed on those points. Toward the end, he said, "We moved one step forward, which is that different countries make different kinds of commitments . . . but that once the commitments are made by developing countries, they need to implement their commitments,"—a decent précis of the day.

I didn't leave the six hours of those talks with Xie thinking that something new or different had happened. I thought I had made the points I wanted to make and held the ground I wanted to hold. But in retrospect, I think those two days together—Xie proudly showing me his hometown, the degree to which my comments in Michigan bothered him, and the intensity and insistence of our back and forth—started to signal a different kind of Chinese desire to put things on a more even keel between us. Not a shifting of the substantive ground China stood on, but some evidence of a desire to find a mutually acceptable path forward with the United States on climate change.

We met again eleven days later in Grenada, on the margins of a ministerial meeting of small island states. He struck a constructive tone about Cancún right away and talked about the importance of continuing to work together toward the South African COP in 2011. I asked whether he envisioned a legally binding agreement there. He said he did, and I repeated, as I'd done so many times before, that we could not do that unless the agreement was legally binding for at least the major developing countries as well. I added that we wouldn't feel the slightest political pressure at home to enter into a binding agreement if China and other majors were not bound.

His response was interesting: "As I understand it, you are showing your bottom line." This was a phrase of appreciation in Xie's lexicon, conveying a sense that I had showed him trust in sharing my assessment with him. As I learned over the years, his conception of negotiating was to push each other until we each understood the other's bottom line, at which point we would be able to work out a deal. He said he would reciprocate by showing me his bottom line, though he didn't really, other than to say—perhaps referring back to our discussion of my University of Michigan speech—that China would not take any steps backward.

* * *

On November 4–5, again in Mexico City, Mexico hosted the annual pre-COP meeting for around fifty ministers. The meeting got off to a relatively positive start compared to the concerns we had been hearing during the fall. Foreign Minister Espinosa promised to maintain open lines of communication in Cancún, specifically referencing the ALBA countries, the small islands, Africa, and the LDCs. Costa Rica, on behalf of the new Cartagena Group, promised "to become more vocal heroes of compromise." Algeria, speaking for the Africa Group, noted a "an atmosphere of renewed confidence" and declared that the commitments taken in Copenhagen should be confirmed.

The most important development at the pre-COP was that India's Ramesh seized the initiative on transparency. At breakfast on November 4, he handed me a single page of ten bullet points meant to sketch out in simple terms the operational elements of a transparency system. These included an understanding that the transparency process would be facilitative and not judge the adequacy of a country's basic pledge; the frequency with which countries would submit a transparency report; the basic content of such a report, including an emissions inventory, description of all mitigation actions, analysis of the impact of those actions, methodologies and assumptions underlying a country's pledge, progress made on achieving a country's pledge; and a consultation session for parties during one of the annual UNFCCC gatherings so the parties could discuss the transparency report of the country in question.[5] Once again, Ramesh had seized the bull by the horns. I told him I thought his paper was excellent, but that he had omitted one important piece: an expert panel to review country reports. He agreed and added it to the next version. Thereafter, Ramesh's plan became a focal point for transparency discussion and debate.

At one point during the pre-COP, the Mexicans engineered a short meeting about transparency for Ramesh, Minister Xie, Brazil's Luiz Figueiredo, and me, with New Zealand minister Tim Groser acting as a "facilitator." It took place in a small room with no table or chairs that looked like the backstage of a theater. Figueiredo talked about how we couldn't have a system that judged or criticized the adequacy of a country's program—Brazil's consistent anxiety. Xie stuck to his premise that Cancún should address only high-level principles—that the system be facilitative, nonpunitive, and nonintrusive—leaving the "details" for next year. I answered that we could certainly do that, but then would have to do the same thing on finance and technology. Figueiredo argued that we should avoid the consultation session with other countries of the kind Ramesh had proposed, but Ramesh answered that you can't do international consultations and analysis without consultations, and you can't do consultations by email.

Toward the end of the first day of the pre-COP, Ramesh made a sly point. He praised China for taking part in 2007 in a detailed environmental performance review, complete with fifty-one recommendations, carried out by the Organization for Economic Cooperation and Development (OECD), an economic body that includes most developed countries. He said he found it so useful that he was going to ask the OECD to do something similar for India. Ramesh's unstated point was clear: if China can do *that*, surely it can take part in the transparency system that Ramesh had outlined.

Two weeks later, at a MEF meeting we hosted near Washington, a range of countries including Singapore, Korea, South Africa, Colombia, Barbados, and Mexico expressed support for Ramesh's ten-point transparency proposal. Ramesh participated by video link from India, sounding like a walking Nike ad, answering those who claimed there wasn't enough time to negotiate operational provisions in Cancún by saying, "We have the time. Let's just do it."

The Cancún COP

COP 16 began on November 29, and its theme might just as well have been "what a difference a year makes." There were real challenges, but not the churning anxiety of COP 15. The Mexicans' guiding principle all year had been to build confidence as well as guarantee inclusion and transparency. No one would be left out, everyone would be welcome to meetings, and there would be no Mexican text. Everything about Cancún was welcoming. The delegates stayed and worked in the same large, spacious, informal resort called the Moon Palace, without the usual headache of transport to and from a convention center. If Copenhagen was cold and gray with wet snow, unhappy crowds, endless lines to get into the chilly Convention center, and lousy food when you got there, Cancún was the opposite. The air was soft and warm, the sky was blue, the Gulf of Mexico lapped the edges of the beach, and the food was good and plentiful. Superficial? Yes. But human nature is what it is, and the Mexicans had not missed a trick.

Two issues stood out for us: securing good language on transparency that was consistent with and went beyond the spare international consultations and analysis formulation from Copenhagen, and anchoring the post-Copenhagen emission pledges made by developed and developing countries in a formal decision adopted by the Cancún COP. On Monday, December 6, the start of the COP's second week, the Mexicans named facilitators to find broadly acceptable solutions on the issues where negotiators were stuck, including Tim Groser to manage transparency. He was a good choice. Short, with white hair and black glasses, Tim was a no-nonsense, experienced, multilateral negotiator for New Zealand who had cut his teeth in the world of trade. He was sharp, irascible, impatient, proud, funny, opinionated, always strategic, instinctively pragmatic, a great conversationalist, and an admirer of the United States. He set to work right away, meeting with us, the Chinese, Indians, Brazilians, and others.

* * *

Thursday, December 9, was the pivotal day of the conference. At around 1:00 p.m., I joined a group of some fifteen negotiators and ministers whom de Alba had summoned to a private suite, Fiesta 2100, away from the main buildings that housed the meeting rooms and plenary halls. China, India, Brazil, and South Africa were there as well as the European Union, United Kingdom, Germany, Japan, Australia, Russia, Colombia, Venezuela, and the Marshall Islands. Brazil's Figueiredo and the United Kingdom's Chris Huhne facilitated, while de Alba presided. The purpose of the meeting was to cut the Gordian knot that had so far prevented agreement about how to anchor the post-Copenhagen mitigation pledges of developed and developing countries in a formal COP decision adopted in Cancún. The reason incorporating post-Copenhagen pledges was so complicated is that it required a double balancing act in which no one on either side felt that they were being disadvantaged. Developing countries would resist a formulation that seemed at odds with their core attachment to the principle of differentiation, but developed countries, the United States chief among them, would resist a formulation that seemed to backslide from the parallel, less differentiated structure of the Copenhagen Accord. At the same time, proponents of the Kyoto Protocol would resist any formulation they perceived as undermining their core objective of securing a second commitment period, to begin after 2012, but countries like Japan, Russia, and Canada, which had made it clear they were finished with Kyoto, would resist any formulation they perceived as implying that they might participate in a second period after all. So the task at hand was delicate.

De Alba's unstated but clear message was that no one was leaving Fiesta 2100 until we had accomplished our mission. We left twelve hours later.

As we were getting settled at the start of the meeting, my English friend Pete Betts looked at me with concern, and whispered, "Where's Sue?" Now Pete and Sue had been sniping at each other for days in what seemed like a downward spiral of mutual pique. Both were tough, blunt, and in a particular frame of mind, gave no quarter. And both had been in that frame of mind all week. But at this moment of truth, Pete knew that if there was one person in Cancún you most wanted in the room to help navigate narrow straits, it was Sue, who can do more with a handful of words or the judicious use of the passive voice than most anyone on the planet. Of course, I knew she was on her way and assured Pete that we could begin.

For a long time, the conversation went around and around. Countries on opposite sides of the double balancing act repeated their mutually exclusive positions, insisting on words the other side couldn't accept. Huhne annoyed more than a few participants because he pushed for UK and EU positions instead of playing his assigned role as a facilitator. At one point, Venezuela's histrionic Claudia Salerno, famous for her bloody hand performance in the final plenary in Copenhagen, had all she could take and stormed out of the room (de Alba knew better than to try to stop her), returning an hour or two later. As we were on a break, around 9:00 p.m. or so, Karsten Sach, a rangy, able negotiator from Germany, and Japan's Shinsuke Sugiyama unearthed some Mexican tequila in a cabinet along with paper cups and poured drinks all around, though some of us, Sue and I included, declined, preferring to keep 100 percent of our wits about us.

Eventually the Rubik's cube was solved in two ways. First, with pivotal input from Sue, two nearly identical sentences were fashioned—each thirty-one delicate words long—to be dropped into the developed and developing country sections of the Cancún decision. Each sentence takes note of mitigation actions to be implemented by the parties, and neither the developed nor developing countries in the meeting believed their preference (more differentiation versus less) was getting short shrift. Second, the Kyoto Protocol problem was avoided with some in-the-weeds sleight of hand (another Sue special) that involved developed country Kyoto parties listing their Copenhagen pledges under both the UN Framework Convention and

Kyoto Protocol. That way, developing countries didn't have to worry that the Kyoto Protocol was being slighted, while developed countries that had decided not to take part in a second Kyoto period were also listing under the UNFCCC, and so didn't have to worry that their actions implied acceptance of a new Kyoto period.

In the meantime, while we were cordoned off in our suite, Tim continued to consult on transparency, including with Jonathan and Jeff Klein, one of the young lawyers on our team, whose mild manner belied a hard-nosed readiness to go to the mat. By Friday morning, Tim, with his assistants, had produced language he thought all the key players could accept and delivered it to the Mexicans.

* * *

At around 5:00 p.m., the Mexicans distributed the text of what they called the Cancún Agreements. It was excellent, from our vantage point. Pledges had been anchored. In the transparency section, concise paragraphs called on developing countries to do most of what Ramesh had laid out in his ten-point plan. The agreements also included a new formulation on equity, declaring that countries should have "equitable access to sustainable development," another Ramesh creation. Developing countries had pushed all year to make the Framework Convention's principle of equity operational, such as by determining how much "carbon space" each country should be permitted to use, based on how much CO_2 they had already emitted. But negotiating a burden-sharing arrangement like that was undoable because countries would never be able to agree on who was required to do what.[6] After Christiana, the new UNFCCC executive secretary, tracked Sue and me down in the plenary hall on the last day and told us we needed to bless the "equitable access" language to get a deal done, we agreed. We did not love the new formulation, but could accept it since it did not prescribe burden sharing.

When the plenary convened in the evening and Minister Espinosa walked in, Jonathan and I, by design, jumped up to start a standing ovation, though it was clear we weren't the only ones with that idea. We had lived through Copenhagen and knew that the spirit in a hall matters. There was one, angry dissenter—Pablo Solon, Bolivia's dramatic UN ambassador—who called for the floor, denounced the text, and demanded that the agreement not go forward because there was no consensus. I began glancing around nervously to see whether anyone else, such as Saudi Arabia's Al Sabban, was going to join Bolivia. But no one spoke up. The Mexicans had clearly done their homework, reaching out at senior levels to potentially problematic countries. Espinosa, for her part, was ready for Solon, having been warned by de Alba about the threat of this kind of disruption. She listened to Solon patiently and then answered that everyone has a right to be heard but not a right to veto what 193 others want to do. Although her statement was not quite on all fours with the consensus practice of the COP, the hall was thoroughly with her, and she gaveled the agreement through to rousing applause.[7] The mood in the hall was celebratory, one part jubilation, one part relief. Most of all, the UNFCCC had demonstrated that it could function. The dread of failure that hung all year like a shadow over the negotiations lifted. As Ramesh said, "The multilateral climate process lives to fight another day."[8] That was reason enough for celebration.

When the main proceedings were all over, I found myself wandering through the late-night plenary hall down the broad aisles between the wide rows of seats to see friends, colleagues, and even adversaries to share the moment with a shaking of hands, pat on the back, and word of appreciation until the next time we'd see each other, a few months down the road. As I walked on, I saw Bo standing behind Denmark's placard in the middle of a long row of seats, too far away for me to reach. So I stopped for a moment where I was, and we looked at each other for a long time, exchanging rueful smiles, thinking the same thing, about the rich irony all around us. Denmark had been excoriated all year for the supposed sin of COP 15, and yet what were the Cancún Agreements if not the Copenhagen Accord expanded from two and a half pages to 30? Yet the Danes were scorned, while the delegates of COP 16 were on their feet cheering for Mexico. We shrugged a "what can you say?" shrug, and then I moved on down the aisle.

Postmortem

The general takeaway in the press and among observers was that Cancún was a modest success. Its achievements did not break new ground, it did not purport to be the new treaty people had hoped for in Copenhagen, and hard issues such as what to do about Kyoto or a potential future legal agreement remained unanswered. But this assessment missed something important. The Copenhagen Accord was a pivotal instrument, and 2010 was its year of maximum vulnerability because those determined to preserve the firewall would try to snuff it out in the cradle. So the stakes for Cancún were quite high, and its success no small matter. The elements of the accord that started to shake the foundation of the old negotiating order were now embedded in an official decision and opened the door on a future yet to be defined.

A number of factors were crucial in Cancún's success. The Copenhagen Accord itself gave the 2010 negotiations a scaffolding on which to build. The Mexican team, led by Minister Espinosa and de Alba, with the backing of President Calderón, ran the right process for the post-Copenhagen moment, calming jangled nerves, assuring even the most cantankerous that they would be heard, delivering the openness and transparency they promised, and circling the globe in a skillful, tireless diplomatic effort. India's Ramesh played a singular role all year in advocating for and delivering operational language on transparency—a make-or-break issue for COP 16. The new Cartagena Group of progressive developing countries combined with the United Kingdom, Europe, Australia, and New Zealand made a difference. It started to free developing countries that wanted ambitious outcomes from the lockstep demands of the G77. As Pete said years later, "We changed the balance of power."

The United States played a crucial role in many ways, but especially in our rock-solid insistence that all the elements of the balanced package had to move together. This forced countries that cared about the assistance elements of the deal to press for a good outcome on the mitigation and transparency issues. China, as always, was important. Premier Wen had played a key role in the end in Copenhagen as well, but this was different. Minister Xie might have been pleased with a more minimal outcome, yet as I had sensed—or at least hoped—in our series of fall bilats, he proved interested in finding a modus vivendi with the United States, which in turn led him toward a result we could both accept.

Finally, the fear countries felt about the impact of failure on multilateralism influenced their readiness to accept compromise. And so another year was over with more progress made. It all came a lot slower than the magnitude of the problem demanded. But we were, at least, moving.

5 A Mandate in Durban

During the first days of 2011, my old friend Jim Steinberg, our brilliant, prescient, acerbic deputy secretary of state, asked me a straightforward question as we sat on couches in his capacious, old seventh floor office. Steinberg thought Cancún had gone well, and a few months later publicly praised the Copenhagen/Cancún "model."[1] What he wanted to know on January 7, though, was what my objective was for 2011 beyond finishing up the unfinished business of Cancún. But completing Cancún's unfinished business was exactly what Sue, Jonathan, and I had in mind for COP 17 in Durban. We saw Durban as something like the third part in a trilogy following on Copenhagen and Cancún: a time to negotiate the guidelines for transparency, set up a Green Climate Fund, flesh out the new Climate Technology Centre and Network, and in general establish an operational system for the decade of the 2010s. Of course, we were well aware that many countries wanted to use the Durban COP as a launching pad for a new legally binding agreement, but owing to the formidable Senate barrier the United States faces when it negotiates a binding international accord, we hadn't embraced the idea.

At any rate, I gave an improvised, unsatisfying answer to Steinberg's beyond- Cancún question, and Steinberg—thin like the long-distance runner he is, with a slight slouch, skeptical eye, and little patience—shrugged, unimpressed. He said he thought it unsustainable for the United States to suggest in our conduct that Cancún was all there was, the full measure of US ambition and intention. Countries were looking for more. Our posture couldn't be "the rest is commentary."

Of course, we were not wrong about the need to complete the important first steps that Copenhagen and Cancún had taken to move past the old paradigm of the climate negotiations. But my conversation with Steinberg, in retrospect, showed two things: that I uncharacteristically misjudged where the negotiations were headed in 2011 and was not yet thinking clearly enough about where the international effort needed to go in the next several years to have real impact. Another brief conversation a few months later echoed my talk with Steinberg. Shortly after giving a speech at MIT on April 21, I sat down with a small group of MIT professors, and one of them, John Reilly, asked, again, a simple enough question: What was our preferred path going forward on the international front? I heard myself giving an inadequate answer, talking about US equities, but not articulating an ambitious vision for where we wanted to go. A month later, in May, I wrote myself a frustrated note about being too preoccupied with managing our way through the COPs without damage. My larger objective had always been to establish as ambitious an international climate regime as possible under a new paradigm that would allow all countries to join. Yet as my inadequate answers to Steinberg and Reilly made clear, in the first months of 2011, we were playing too much defense, not enough offense. That would have to change.

Kyoto and a New Mandate

The climate negotiations in 2011 were shaped by three essential realities. First, the Kyoto Protocol's initial commitment period, running from 2008 to 2012, was coming to an end, so postponing a decision on Kyoto for one more year was no longer an option. Second, developing countries fervently wanted a second commitment period of Kyoto to be agreed on—some, including China and its allies, because they saw preserving Kyoto as a means of protecting the firewall, at least until 2020, and others, including the islands, LDCs, and progressive Latins, because they believed its rules-based, legally binding approach would serve as a model for a new legally binding agreement. Third, the Europeans were the only ones who could keep Kyoto alive since the United States had never joined and other significant Kyoto parties, including Japan, Canada, and Russia, had declared they would not participate in a second commitment period.

These Kyoto facts of life meant that the Europeans had the leverage in 2011 since they were the only ones who could deliver what developing countries so ardently wanted. And the European Union intended to use that leverage, saying early and often that its price for recommitting to Kyoto was a new mandate for a legally binding agreement. Whether it could use its leverage to achieve that end or something close to it remained to be seen. Leverage only works if others believe you're prepared to carry out your implicit or explicit threat—in this case, for Europe to refuse a second Kyoto commitment period if it could not secure the mandate it wanted in Durban. And some were skeptical for a variety of reasons that Europe would ever really walk away from Kyoto; many in Europe saw Kyoto's rules as valuable in considering what a new agreement should look like, were sensitive to the views of their green public, and would have been loath to abandon Kyoto before a new legal agreement was at least in the works. In addition, unlike both the United States and China, the Europeans did not have a reputation for playing hardball.

* * *

Our own concerns about a new mandate derived from the US hurdle imposed by the Constitution's Treaty Clause, requiring the approval of treaties by two-thirds of the Senate.[2] No other country labors under this kind of burden. At the same time, not every international agreement is a "treaty" requiring Senate "advice and consent." In fact, a great many international agreements that the United States enters into year in and year out are not considered treaties as the United States understands that term. And whether or not an agreement is considered a treaty for US purposes depends on the language of the agreement—what it requires, how it is worded, and whether there are other decisive factors in play, such as that a given obligation can be fairly construed as an extension of provisions in an agreement already approved by the Senate—such as the 1992 Framework Convention.[3] We knew that which side of the line a new climate agreement would fall on—requiring Senate approval or not—would likely determine whether the United States would be able to join. We knew this because the road to Senate approval of international agreements had, by the time of the Obama administration, become treacherous. The 1982 Law of the Sea Convention, adhered to by some 168 countries, but not the United States, was modified in 1994 to address Reagan era objections about deep seabed mining. Those US-inspired fixes, coupled with strong support from the US military and all relevant US industries, should have paved the way for easy Senate approval. Yet concerted efforts to secure such approval first by President George W. Bush and then by President Obama failed, even though ratification would have benefited the United States in myriad ways. Or consider the 2006 Convention on the Rights of Persons with Disabilities, inspired by our own Americans with Disabilities Act of 1990. Such lineage wasn't good enough for the Senate's "black helicopter" caucus, which blocked its approval.[4]

* * *

Discussions leading to Durban went on in earnest from April through November, and often had the character of a three-cornered game. The European Union envisioned a legally binding agreement that would apply to all countries. Mexico's de Alba offered one of the more insightful suggestions of the year at our April MEF, when he responded to EU commissioner Hedgaard's call for a legally binding agreement, "My main point is that we need to depart from a 'yes or no' debate. It is not helpful. Rather, why do we need it? What topics may be needed? I think . . . a combination of soft law [urging or encouraging rather than requiring] and hard law may well fit our needs in the immediate and long term." At our November MEF meeting, de Alba returned to this point and said, "Mexico's preference is to look into a regime in which binding and nonbinding commitments will combine. It is not a yes or no system." I thought de Alba was entirely right. Indeed, two years later, I made a speech at Chatham House in London in which I talked about the importance of soft as well as hard law.

The United States was never an advocate for a legal mandate, but during the course of the year made it clear that we could support a mandate for a legally binding agreement if the mandate called for true legal symmetry as between developed and at least major developing countries, included no "escape hatch" words that would let developing countries off the hook, and left open the question of what elements would be binding in the agreement. In our view, if the United States found itself once again unable to join a new climate agreement, it would be a debacle not only for us but also for the world since it would leave the agreement ineffective, if not stillborn.

Discussions that Jonathan, Sue, and I had with the European Union, United Kingdom, France, Germany, and several others on the sidelines of the ministerial meeting South Africa hosted September 8–9 in Pretoria made us uneasy about how tough the European Union would be in demanding what we would need in a new mandate. The Europeans asked us whether we really needed clarity about legal symmetry in the mandate itself since we could always push for such symmetry when the actual agreement was negotiated. They urged us to let the process "get going" and not demand too much now, but I said no. If we did not insist on symmetry in debate over the mandate, we would be hard-pressed to recapture it in the agreement to be negotiated under that mandate. At our September MEF meeting, Christiana Figueres, the executive secretary of the UNFCCC secretariat, tried out her version of the Europeans' point, saying that while "the devil is in the details" for any proposed mandate, "the beauty of Durban is that it doesn't have to spell out all the details." Except that spelling out certain details is exactly what we needed. There were plenty of ways for developing countries to build an escape hatch into a future agreement, such as a reference to equity or CBDR, or invoking Article 4.7 of the Framework Convention, which many developing countries read to mean (wrongly in our view, but nonetheless) that they didn't need to undertake actions to reduce emissions unless developed countries paid for them. If there was going to be a new mandate, it would have to shut the door on those kinds of evasion.

As for China, our interests in 2011 narrowly overlapped in the sense that neither of us wanted a mandate for a legally binding agreement. But otherwise, we were mostly at odds. Minister Xie made his orientation clear at our April MEF meeting, saying, "We have the principle of common but differentiated responsibilities. We have a comprehensive foundation—the Convention, the Kyoto Protocol and the Bali road map. These are the existing legal basis. . . . If there are imperfections, we can improve it. . . . We cannot go back. . . . So we insist on working on the existing framework." By invoking what he saw as the three foundational documents of climate change—his holy trinity—he was asserting the preeminent role of the firewall as if Copenhagen and Cancún had never happened.

He also advocated a go-slow approach, arguing at our November MEF meeting that countries should wait for completion of the 2013–2015 review called for in the Cancún Agreements (to assess the adequacy of Cancún's long-term global temperature goal) and then "we can enter into the phase of discussion and negotiation to decide what we need to resolve for the period after 2020." The European Union's Hedegaard hit the go-slow proponents hard: "Can you imagine what it would look like if we had to communicate that it took another three or four years to think, reflect, assess things?" De Alba emphatically agreed.

* * *

The consequential event of the year in India, from our perspective, was that Ramesh was part of a July government reshuffle that sent him to the Ministry for Rural Development. Days before this was announced, at the early July Petersberg Dialogue meeting in Berlin, Ramesh delivered one more provocative idea to the climate negotiations. He suggested that we all needed to start rethinking what we mean by legally binding and be more flexible. He contended that legal bindingness at the domestic level is more meaningful than at the international level, since domestically binding laws and regulations are enforceable while internationally binding climate targets would certainly not be, given that too many countries would oppose enforcement provisions. He also noted that actions taken at the national level could, in effect, be brought into the international arena by way of an effective transparency regime. Of course, we liked his idea, which was similar to one Sue and I had included almost two years earlier, in an August 17, 2009, memo we did for the White House. Regrettably, that was Ramesh's last contribution to the climate debate. His intellect, creativity, willingness to step outside conventional channels, and fearlessness along with the joy he took in all of it was sorely missed.

The Durban COP

I arrived in Durban with Sue late Friday night, December 2, and spent my first day meeting with my team, the Umbrella Group (non-EU developed countries), the LDCs, AOSIS, Tim Groser (New Zealand), André do Lago (Brazil), and the European Union's lead negotiator, Artur Runge-Metzger. The most aggravating discovery of the day was that virtually all country delegation offices, including ours, had been set up in the claustrophobic, underground parking lot of the International Convention Center, though by midafternoon I'd found a work-around, setting myself up at a table in the lovely outdoor courtyard adjacent to what was about to become my new favorite coffee and sandwich stand. The courtyard became my working office for the next seven days.

* * *

In my meeting with Artur, he told me that developing countries were demanding not just that EU countries agree in a simple COP decision to carry out a second round of Kyoto targets but also that they execute a formal amendment to Kyoto too, necessitating the unwelcome, laborious process of ratification by each of the European Union's twenty-eight member states.[5] This development only strengthened the European Union's resolve to insist on a mandate for a future legally binding agreement as the price for its adherence to Kyoto. I repeated my catechism that the United States could accept a new binding agreement only if China, India, and other majors did so as well, but that they were plainly unwilling to do so. Artur, a first-rate negotiator who is slim, shrewd, bespectacled, and doles out his words carefully, gave me his best Cheshire cat grin and said, cryptically, "We shall see." This made me nervous.

In particular, it triggered my old concern that the European Union would demand legal symmetry in the mandate for a future agreement, but then allow the BASIC countries and their allies weasel words they would cite to avoid legally binding treatment for themselves.[6] In a note I wrote in late October, I envisioned the three-cornered contest between the European Union, United States, and BASIC countries as a game of musical chairs. The European Union might be left standing, forced by the combination of the United States and BASIC to accept a weaker mandate than it wanted—one that did not ensure a legally binding form. Or the United States might be left standing, as the European Union and BASIC collaborated on a legal mandate that allowed developing countries an artful dodge, reflecting only a faux symmetry. I did not doubt the European Union's desire for a new legal agreement that applied to all the major developing countries, but I did worry that as the pressure built toward the end of the COP, the European Union would give BASIC an out in order to get a mandate done. In my imagined game, I did not see the BASIC countries left standing. But this turned out to be closer to what happened in the end.

The three-cornered game produced some strange bedfellows in Durban. In the overall struggle to contain climate change, the United States and European Union were always most strongly aligned, despite some differences. We shared a conviction about the need for ambitious action and agreed that all significant emitters had to be part of a single, nonbifurcated system, beyond the old firewall. But the European Union was also devoted to a legally binding approach that would require Senate approval, thus leaving the United States out of the agreement—an untenable position for us and the world as well. So we found ourselves in Durban collaborating with the BASIC countries to avoid legally binding language at the same time that we were working with the European Union to make sure that whatever legal language was agreed on did not include escape hatches and would not require Senate approval.

* * *

On Sunday, I met with Minister Xie in China's bright, spacious, above-ground quarters—the only country with such pleasant, airy digs. This was our fourth meeting since September, and we had spent a lot of time talking about the potential mandate for a future agreement as well as my concerns that China seemed once again to be retreating from what we had agreed to the year before on transparency. They and their allies had offered up sophistical arguments all year long: the international consultations and analysis of *biennial* reports that the Cancún Agreements required developing countries to undertake (¶63) only needed to be done every *four* years; the detailed content about what developing countries needed to include in their transparency reports set forth in Cancún (¶64) did not need to be included in their reports after all; the facilitative sharing of views that was supposed to take place in the UNFCCC's Subsidiary Body for Implementation (¶63) did not have to happen; and the statement in Cancún that the content and frequency of communications from developing countries "will not be more onerous" than for developed countries (¶60[a]) actually meant that they had to be less onerous. None of this passed the straight-faced test. For us, it underscored why we believed any mandate needed to be clear and precise. We lived in a world of frequent, if dubious, reinterpretation.

On the mandate question, Xie sharpened the argument he had started to make when we met in Beijing in October. There he had attacked my assertion that the United States could not do a legal agreement because China would not, saying, with some justification, that "even if China can do a legal agreement, the United States cannot." In Durban, Xie turned this critique into a warning: "Do not link the US position on legal with China's position. . . . Don't push me hard or I will push you hard." This was uncharacteristic language for Xie—not the way we had ever talked to each other. But it was a sign of the pressure he felt. China certainly did not want to get cornered into accepting a mandate for a new legally binding agreement that would bind them. But Minister Xie undoubtedly saw the strong alliance between the European Union and the vulnerable and progressive countries as disturbing because it threatened to set China up in opposition to those developing countries, precisely the problem China had confronted in Copenhagen. He didn't want China to absorb the blame if the push for a legal mandate failed in Durban. I ignored his threat and the anxiety underlying it, and answered that the United States could not justify a post-2020 legal agreement unless all the major players were in on the same legal footing. Once we started thinking about the world of the 2020s, I thought a firewall became particularly untenable as a matter of climate ambition—quite apart from US political concerns—given rapidly rising developing country emissions.

* * *

Jonathan, Sue, and I had an important dinner with the Brazilians on Monday evening at a small, open-air restaurant called the Ninth Avenue Bistro. Luiz Figueiredo and Andre do Lago were joined by Andre Odenbrite, sometimes referred to as "little Andre" (though he wasn't small) when in the presence of 6′7″ do Lago. Luiz and (big) Andre were by this time good friends of ours. Luiz is a tough, proud, skilled diplomat who exudes self-confidence and reserves a touch of scorn for those who, in his view, aren't quite up to snuff. He knew the issues inside out, knew the players, and was an adept tactician. He ended up being an important player all the way to Paris, having moved through very high-level positions, as Brazil's ambassador to the United Nations, foreign minister, and ambassador to the United States. I first met him on a trip to Brazil in 2009, and that same year we began our annual dinners in New York with Jonathan Pershing at Avra, a Greek seafood restaurant, during the UN General Assembly meetings. Andre is a born diplomat of charm, grace, intelligence, and wit, leaning over in the way the very tall do so as not to miss any words, managing to be entirely committed to climate action and serious about his mission while keeping a fine eye on the inevitable absurdities of the game. At a pre-COP meeting the next year in Seoul, Sue, Jonathan, and I had joined the Brazilians for an impromptu chat, sitting on couches in the spacious lobby of our hotel. At one point, I took the opportunity, in a "you've got to be kidding me" tone, to chastise our friends about their recent efforts to introduce CBDR into all sorts of nonclimate negotiations. Looking at him, I said, "Andre! I mean come on, the Universal Postal Union!" And he rocked back, clapped his large hands together in delight, and burst out laughing.

What was clear from our meetings in the courtyard and at dinner was that Brazil had come to Durban with a mission to secure a successful outcome. Earlier in the year, Luiz had talked down the notion of a legal mandate, arguing that to decide on a legal form before we knew the content was like signing a blank check. But in Durban, Brazil had turned a page. Six months after the COP, it would be hosting a major event—the Rio + 20 Conference on sustainable development—commemorating the original 1992 Rio Earth Summit where the Framework Convention, among other agreements, was completed. Luiz would be managing the conference with Andre serving as his chief negotiator. So Brazil was keen on success in Durban in part because it did not want the detritus of a failed COP to wash up on the shores of Rio and become its problem to fix. Brazil was also keen to benefit from the firewall protection of the Kyoto Protocol until at least 2020. More broadly, Brazil feared that if the Durban conference failed, it could lead to the broader failure of the Framework Convention itself—the fundamental guarantor of fair treatment in Brazil's view and the anchor of any future international climate regime. And it was easy enough to see how that failure could happen given the kind of checkmate logic I had articulated at our September MEF meeting: the European Union would only agree to a Kyoto second commitment period if there was a new mandate for a legally binding agreement; the United States would consider such a mandate only if it was equally binding for at least the major developing countries; but China and India, among others, would not accept such symmetry.

So the Brazilians, always deft strategists and tacticians, came to Durban, with the blessing of President Dilma Rousseff, Foreign Minister Antonio Patriota, and Environment Minister Izabella Teixeira, determined to help bring the COP in for a safe landing. They recognized the large-scale support for a future legal agreement among both developed and developing countries. They talked to China and India about the importance of working constructively toward a new agreement rather than trying to block it, and sought to persuade them that if the G77 secured a second commitment period of Kyoto, China and India would have to accept a new mandate.

The focus of our dinner and ongoing conversations the rest of that week was to find language that went far enough down the legal track that the European Union and its developing countries allies could accept it, but not so far that the United States or BASIC countries could not. I reminded Luiz and Andre that if their side were to secure mandate language that was soft and qualified for BASIC but clear and binding for the United States, it would be a pyrrhic victory for them since the United States would be unable to join. Throughout the rest of the conference, we stayed in close contact with Brazil.

* * *

On Wednesday, December 7, less than three days before the scheduled end of the COP, a more intensive effort took shape to hammer out a mandate for a future agreement that all parties could accept. In particular, at Brazil's suggestion, South Africa set up a meeting of the European Union, the United States, China, India, and Brazil to seek common ground on the mandate. The meeting took place off-site in the Hilton Hotel adjacent to the conference center. It started at 5:30 p.m., broke for dinner at 7:00 p.m., resumed from 9:30 p.m. to 12:15 a.m., and then again the next afternoon. South Africa participated in its role as COP president. We talked about mandate language describing the legal nature of the agreement, the time frame to start and finish negotiations for a new agreement, and the perennial question of differentiation. On the legal characterization of a new agreement, the European Union favored saying the new agreement would be legally binding, but the United States and BASIC countries objected.

At some point in the Durban discussions, the legal provision in the proposed mandate took the shape of a sentence with options. Rather than calling for a legally binding agreement, the mandate would call for a "protocol or another legal instrument." On Thursday, December 8, South Africa held another of its "Indabas," a Zulu and Xhosa word for a meeting of a community's wise ones to discuss difficult problems. Around seventy countries sat around a large rectangular table, and during that meeting, I announced that the United States would accept the phrase "another legal instrument. By this time, it was obvious to us that Durban would not be successful unless the United States found a way to support an explicit reference to a new legal agreement, and we concluded that "a protocol or another legal instrument" met our bottom-line need for flexibility regarding which elements in a new agreement would be legally binding. Pete Betts told me later that he thought our move to accept that legal formulation helped move the negotiation forward.

The trilateral meeting also discussed whether a new agreement would reference the classic principle of CBDR. In our view, such a reference, unless qualified, would create the precise asymmetry in legal force between developed and developing countries that we could not tolerate. For that reason, throughout the conference, I had said that the United States would accept a reference to CBDR in the new mandate only if it was modified to capture the idea of evolution over time—CBDR as appropriate to the time period in question, whether the 2020s, 2030s, 2040s, or beyond. The BASIC countries, in particular China and India, flatly refused to consider any new CBDR formula suggestive of evolution, insisting that CBDR be applied on behalf of all countries classified as developing (non-Annex 1) in 1992, irrespective of how much they had grown and developed since then. China was especially adamant about blocking evolution language, given its galloping industrialization and status as the world's second-leading economy.

* * *

Overall, the Thursday night Indaba did not make a great deal of progress. Shortly after midnight, Foreign Minister Maite Nkoana-Mashabane announced that she was adjourning it, but Connie Hedegaard and I along with several other delegates prevailed upon her not to adjourn since we had only one more scheduled day in the conference. Mashabane then announced that any countries that wished to continue discussions should go to a small conference room, called Sobistar, with space for around thirty people to sit around its square table. The conversations there continued until around 4:00 a.m. During the course of the next two days, much of the negotiating action took place in two additional Sobistar sessions, chaired by South Africa, the first on Friday night until around 2:30 a.m., and the second on Saturday afternoon.

The central players in the Sobistar discussions were the other BASIC countries, China, India and Brazil, the United States, European Union, and Cartagena countries, including Colombia (Andrea Guerrero, playing a key role), Barbados (Selwin Hart), Grenada, and Gambia (Pa Ousman). The core language on the legal character of a new agreement referred to a new "process to develop a protocol or another legal instrument." The symmetry of the mandate was largely established by the inclusion of a phrase declaring that the new agreement would be "applicable to all parties"—effectively, the opposite of Kyoto.[7] But this symmetry could still be unwound by the wrong words on common but differentiated responsibilities. The BASIC countries made another hard push to include CBDR language, but I stuck to our Durban position that we would accept a CBDR reference only if it made clear that CBDR was an evolving concept. China and India said no, so we were at an impasse. During a break in the action in the Sobistar room on Saturday, Luiz Figueiredo ambled over to me and Sue and quietly suggested that we leave any reference to CBDR out of the mandate but acknowledge that the new agreement would be "under the Convention," an idea he first floated with us at our dinner on Monday night.

This was another example of the constructive ambiguity that allows difficult negotiations to conclude, but can also lead to friction and discord in the months and years following. We said yes to Luiz's suggestion because we looked at "under the Convention" as stating the obvious—no one doubted that a new agreement would be an offspring of the Framework Convention, meant to carry out the Convention's core objective of preventing dangerous climate change. But from Luiz's perspective, on behalf of BASIC, the principle of CBDR is part of the Convention and thus could be argued to have come into the Durban mandate through the back door. Still, the constructive ambiguity of a phrase does not mean that the two sides have simply split the difference. We got the better of this deal because "common but differentiated responsibilities," always seen as vitally important for those who defended the firewall, was conspicuously absent from the mandate, and the back door is still not the front door.

On timing, the BASIC countries were determined to stay under the protective canopy of the Kyoto firewall through 2020, so wanted no new agreement to take effect before then. They also pushed to delay negotiations until 2015, but had little support for this temporizing approach. Developing country progressives in the Sobistar room wanted to start negotiations right away, finish within a few years, and produce a legally binding agreement taking effect by roughly 2015. EU countries supported their Cartagena allies but were more focused on having a new agreement completed by 2015 than on having it actually take effect by then. The European Union knew that getting a new agreement not just negotiated but also entered into force by 2015 would be a tall order, yet they saw concluding the negotiations by 2015 as crucial for the simple reason that the December 2015 COP would be the last one in which President Obama would still be in full power (assuming, of course, that he was reelected in 2012). A December

2016 COP would take place in Obama's lame-duck phase, with a new president already elected. As things turned out, with Trump succeeding Obama, the European Union proved to be more prescient than it knew. As for us, we were comfortable with the European Union's timeline and knew that the BASIC countries would not accept any new start date before 2020. The compromise reached in the Sobistar discussion called for negotiations to begin in 2012 and finish by 2015, but for a new agreement to take effect "from 2020."

The Cartagena progressive and vulnerable countries played an important role in securing strong language in the mandate's preamble, calling on "the widest possible cooperation by all countries" in accelerating action to reduce emissions. In addition, AOSIS, the island group, succeeded in its push to strengthen language regarding the global temperature goal, describing the desired emission pathway as consistent with a likely chance of holding the increase in global average temperature "below 2°C or 1.5°C," the first time 1.5°C had elbowed its way into the main temperature goal. The BASIC countries and their allies did not like either of these additions, the first because it applied to "all countries" rather than demanding more action by developed countries, and the second because the Chinese in particular had never been a friend of the 1.5°C target, recognizing the pressure it would put on them to cut their emissions more aggressively.

The Sobistar discussions also agreed that the Bali mandate and the "LCA" ("Long-term Cooperative Action") working group established under it would be terminated in 2012, following COP 18. This was a priority for us because we were not prepared to allow the Bali mandate—interpreted by so many as perpetuating the firewall—to influence any part of the new negotiation. China's concerted effort earlier in the year to assert the primacy of Bali in order to protect the firewall had only sharpened our determination to have the Bali mandate formally closed. The Durban mandate established a new negotiating track called the Ad Hoc Working Group on the Durban Platform for Enhanced Action (ADP).

* * *

Meanwhile, negotiations to implement the relevant elements of the Cancún Agreements had concluded in decent shape, including guidelines on transparency, the formal launch of the new Green Climate Fund, the establishment of a new standing committee to assess climate finance flows to developing countries, and implementing decisions for both the new Climate Technology Centre and Network and the new Adaptation Committee.

* * *

Although things seemed settled when the last Sobistar meeting broke up, Durban had a surprise in store. India was unhappy with the two legal options included in the mandate in the final Sobistar session ("a protocol or another legal instrument"). It wanted to include a third option, "legal outcome." But Pete Betts, for the EU, had flatly rejected that proposal, disdainfully saying the only thing "legal outcome" guarantees is that the outcome won't be illegal. His concern was that legal outcome could be read to allow a new agreement to be a simple COP decision rather than a true legal agreement, which was precisely the leeway India wanted. In any event, after the Sobistar session ended late Saturday afternoon, India, presumably with China's support, privately persuaded South Africa to add "or a legal outcome" as a third option. Not long after this new deal was struck, Minister Mashabane and her team called Colombia's Andrea Guerrero to their COP presidency basement offices to inform her of the new language and request that she accept it. She refused. The Cartagena members had agreed before the COP that they would not accept an agreement in Durban without a mandate for a legally binding agreement. The protocol or another legal instrument language was already less than they wanted, but Andrea and Colombia's Environment and Sustainable Development Minister Frank Pearl were prepared to accept it as an adequate compromise. But not "a legal outcome," which Andrea scornfully dismissed as "or whatever." The South Africans said Colombia would be alone if they opposed the new language, and told her that the islands and LDCs had already said they would go along with the new proposal.

Incensed, Andrea called Pearl, who in turn called Colombia president Juan Manuel dos Santos. Dos Santos confirmed that Colombia, which had suffered through torrential rains and disastrous floods in 2011, could not accept a weak agreement even if it stood alone. Andrea then found the European Union's Connie Hedegaard and asked, "Why have you abandoned us?" Connie, as fierce and undaunted as Andrea, said she had done no such thing, didn't know what Andrea was talking about, and in the best "this shall not stand" tradition, told Andrea that the European Union would not allow the legal outcome language to pass. So, a potentially fatal plenary collision was set in motion.

When the plenary assembled to consider the new mandate, Connie blasted the new change, declaring that "legal outcome" was unacceptably weak and could put in doubt the European Union's willingness to do a second commitment period of the Kyoto Protocol. Colombia seconded the European Union's protest, saying it could not accept the Durban package as it stood. At that point, Indian minister Jayanthi Natarajan, the best orator in the hall, let loose as only she could, speaking out for equity and declaring that India would "never be intimidated by threats." Minister Xie followed with an impassioned defense of India, shaking his fist and saying to developed countries, "What qualifies you to tell us what to do?" COP president Mashabane, facing a furious brawl that threatened to torpedo the entire COP, asked India, the European Union, and other interested parties to form a huddle and find language acceptable to all.

In the organic way these things happen, the huddle started to take shape and expand, amoeba-like, around Commissioner Hedegaard and Minister Natarajan, growing to some forty people, with Sue and I close to the center as well as Luiz Figueiredo (Brazil), Su Wei (China), and Minister Mashabane. A variety of ideas were proposed and rejected, either by the European Union or India. When someone suggested incorporating the idea of equity to solve the problem, I shook my head "no" since an escape hatch cure like that would be worse than the disease. Finally, Sue—who else?—came up with a near fix. Realizing that there was no adjective before "outcome" that could possibly work for both India and the European Union, she rearranged the phrase and proposed "an outcome with legal force." The Indians listened and didn't say yes, but didn't say no, which was progress. Luiz then managed to slip in through the throng for a quiet word with Natarajan and her team, suggesting that they add the word "agreed" before "outcome." This was a clever stroke because "agreed" made more emphatic what was already true: that no ultimate outcome could occur without consensus, meaning that the Indians could still say no to a new agreement they found unacceptable. In addition, "agreed outcome" was a comforting phrase to Indian ears since the same phrase appeared in the venerated Bali Action Plan of 2007. After a brief pause, Natarajan said yes, "agreed outcome with legal force" was accepted, "legal outcome" was deleted, the plenary reconvened, and COP 17, complete with an approved mandate, now called the Durban Platform, finished its main work. In the end, it was the longest COP to date.[8] Of course, deft wordsmithing alone didn't solve the problem. The moment mattered too, since the weighing and measuring of nuance at the end took place as the entire plenary hall looked on, waiting with bated breath to see whether white smoke or black would emerge from the huddle.

Interestingly, the clever phrase invented by Sue and Luiz had no conventionally recognized meaning in international law. But having "legal force" clearly meant more than simply a "legal outcome" and it might well have meant legally binding, at least in some ways to be determined. All of this would become grist for further negotiations under the new ADP negotiating track.

Postmortem

In the end, the Durban COP turned out to be a crucial pivot for the climate negotiations, delivering a new mandate and new schedule, setting international climate diplomacy on a new path to produce a long-term, operational, legal agreement. The Copenhagen and Cancún COPs as well as the implementation part of Durban did important work, chipping away at a flawed negotiating paradigm as well as putting in place country emission targets, new transparency procedures, and new mechanisms for financial and technology assistance. But it was the mandate hammered out in Durban that heralded the possibility of a breakthrough agreement in the next four years. Elliot Diringer of the Center for Climate and Clean Energy Solutions, consistently one of the most astute climate observers, wrote soon after the COP that Durban "could prove to be a key transitional moment," with "the deal . . . delicately poised between two eras—the fading age of Kyoto, and a new phase beyond Kyoto, with developed and developing countries presumably on a more equal footing."

That Durban ended up this way came as a surprise to many, largely for the same reasons I had articulated during the year: that the European Union, United States, and BASIC countries seemed to have irreconcilable needs. And yet in the end they reconciled, at least enough. For the European Union, the mandate didn't quite spell out a legally binding agreement but it nonetheless spelled out enough that such an agreement was encompassed within what the mandate set in motion. For the United States, the mandate met our needs by being applicable to all, avoiding any escape hatch, and leaving open for negotiation the question of what elements in a new agreement would be legally binding. China and India were the ones dragged furthest from their comfort zones. China, which staked out a position of trying to reject a new mandate and revive Kyoto and the Bali road map, and urged that we did not need to even start thinking about a post-2020 future until sometime after 2015, prevailed in none of these objectives, instead saying yes to a prompt negotiation of a legal agreement applicable to all. India desperately wanted a mandate that could be satisfied by negotiating a nonbinding COP decision, but did not get it. The addition of a third option—agreed outcome with legal force—was better than nothing but not what India desired.

A number of players were instrumental in making the Durban Platform possible. The European Union, together with its allies among the islands, the LDCs, and Latin progressives, heads the list. Pete Betts, who was not only the United Kingdom's director for international climate change but also a lead negotiator for the European Union, considered Durban the high-water mark for EU influence. The European Union used its leverage well, both by refusing to deliver Kyoto until it secured a strong enough mandate and by accepting a mandate that called for a legal agreement but did not specify what elements would be legally binding. And crucially, the European Union worked in close alliance with poor, vulnerable, and progressive countries, which had never before played such a central role. This alliance, with Pete and Andrea Guerrero leading the way, effectively forced China's hand. Ever since positioning itself in opposition to poor and vulnerable countries at the 2009 Copenhagen COP, China was loath to repeat that unhappy experience. In the end, it accepted a genuinely symmetrical legal mandate in important part so that it would not become the antagonist of the poor and vulnerable—though the Durban mandate was unwelcome enough that China still tried to defang it near the end by joining India's "legal outcome" gambit, first quietly, behind the scenes, and then loudly in the open plenary hall.

Brazil played a key role, arriving at the COP primed to use its considerable diplomatic skill to help navigate through a thicket of conflicting objectives and secure a workable outcome. BASIC may be a group of four, but it was already becoming a group of two plus two, with Brazil and South Africa acting as a moderating force, more disposed than China or India to consider new ideas, including, in 2011, a legal agreement applicable to all. The Brazilian team worked closely with us, worked with the South Africans to bring the three big powers (the European Union, BASIC, and United States) together, and privately pushed China and India to go along with a workable mandate, largely on the grounds that it was coming and they would be better off seen as builders of it than having it forced on them.

As for the United States, we were slow on the uptake and dragged uncomfortably into a negotiation for a mandate. But once we joined the debate, we played a crucial role in shaping the mandate so it would continue the necessary shift away from the old, bifurcated negotiating paradigm and define the terms of what became the Paris negotiations.

Countries left Durban on a new path. There was still a long way to go and no real clarity about how to proceed, but after the chaos of Copenhagen and reestablishment of stability in Cancún, Durban opened the door to new possibilities. Jim Steinberg—not surprisingly—had been right back in January. Cancún could not be all there was, and indeed on a quicker timetable than I had originally envisioned, Durban stood as testament to that truth.

6 The Gears Engage

During the next two years, slowly in 2012 and then with gathering momentum in 2013, the scaffolding of what would become the Paris Agreement started to take shape. Our priorities in this period were to defend the Durban mandate against efforts by China, India, and their allies to undermine it through reinterpretation and to start establishing foundational elements for a new agreement. This two-year period was also marked by a visible strengthening of US action at home and intensified effort by the United States to engage constructively with China.

Reinvigorated US Action

By the end of his first term, President Obama saw climate change as an important piece of unfinished business and came out of the gate in 2013 for his second term determined to do more. In his inaugural address, he pledged to respond forcefully to the climate threat lest we "betray our children and future generations." In his State of the Union speech three weeks later, he urged Congress to work on a bipartisan, market-based solution of the kind senators John McCain (R) and Joe Lieberman (D) had proposed several years earlier, and then threw down the gauntlet: "But if Congress won't act soon to protect future generations, I will." Then on a sweltering June day, speaking outside at Georgetown University, Obama made good on that promise with a full-throated call to action centered around a new national Climate Action Plan that would sharply reduce emissions from power plants, double the US use of wind and solar energy, impose tougher fuel standards for trucks and buses, and boost the federal procurement of clean energy. At the beginning of 2014, Obama brought John Podesta back to the White House to oversee a "whole of government" implementation of the new plan. Throughout his next four years in office, Obama never let the climate issue go, staying focused and engaged at home and abroad. All of this activity significantly strengthened our hand in the negotiations.

* * *

Twelve days after Obama's second inauguration, John Kerry, the conscience of the Senate on climate change, began his tenure as secretary of state. Climate change was the issue he cared about most passionately. He was one of the Senate originals on climate, together with Al Gore and Tim Wirth. He was the senator who showed up at the annual COP meetings even if he were a delegation of one. He was a man with a truth, certain he was right, pained that so many failed to see the threat. Of course as secretary of state, he threw himself into the critical issues of the day, from the Iran nuclear negotiations to North Korea to the Middle East. But on climate change, he was lit from within. He had no patience for bureaucratic dawdling or those who failed to grasp the urgency of the moment. Nor did he have much patience for the trench warfare of climate negotiations. He wanted to seize the day, he had a taste for the grand gesture, and yet climate negotiations are, by their nature, a talking game that plays out day by day, month by month, year by year. So the negotiations often frustrated him, but he pushed and prodded and inspired and became a powerful voice for the United States.

Defending Durban

The Durban Platform rattled developing countries most concerned about defending the firewall. In its summary of the first 2012 intersessional meeting in Bonn, May 14–25, *Earth Negotiations Bulletin*, a respected, longtime source of reporting on climate diplomacy, noted that the Platform did not include references to the classic UNFCCC principle of common but differentiated responsibilities, or even draw a distinction between developed and developing countries. "Some parties have started to panic about the ADP [Ad Hoc Working Group on the Durban Platform]; they feel as if they are walking into a dark room and don't know if there is anything there or where anything is."[1]

As an expression of this anxiety, countries most committed to defending the firewall principles of CBDR, equity, and the historical responsibility of developed countries started to form a new group at the May intersessional. It was comprised of China, India, Venezuela, Bolivia, Ecuador, Argentina, the Arab group countries including Saudi Arabia and Egypt, Pakistan and Malaysia, among others. The group came to be known as the Like-Minded Developing Countries (LMDC) and first met formally in Beijing on October 18–19, 2012. Alf Wills, the longtime wise man of the South African climate team, said much later that when the LMDC came together, joined by neither Brazil nor South Africa, it was effectively the end of the BASIC group as a major force. South Africa and Brazil had always been moderating voices within BASIC, and China and India clearly felt they needed a more united hard-line front against what for them was a disappointing outcome in Durban, where Brazil and South Africa had played more accommodating roles. Once it took shape, the LMDC group became that hard-line front.

The most revealing exchange about the firewall occurred at a MEF meeting in New York in September 2013. New Zealand's Jo Tyndall, a skilled climate diplomat, said that the 1992 division of parties into developed (Annex 1) and developing (non-Annex 1) countries could be treated in only one of three ways: deleted, redefined, or left alone, but with no operational significance. She said that the combination of self-differentiation implicit in a bottom-up, nationally determined structure for mitigation, plus flexible, top-down rules for reporting and review, among other things, would differentiate in a nonbifurcated manner consistent with the principles of the Framework Convention. But India's Varad Pande disagreed. He claimed that since nothing had changed since 1992 when the Framework Convention was agreed to, the existing annex system should remain in place:

US chair (Caroline Atkinson): I think a lot of things have changed in twenty-one years.

Pande: "A lot has changed in the world, but the UNFCCC has not!"

Stern: "Yes, that might say it best."

* * *

The instinct among the newly forming LMDC group to defend the firewall above all else crashed into a different impulse among the vulnerable and progressive alliance that first took shape in 2010 as the Cartagena Dialogue and emerged more decisively at the Durban COP. That alliance cared most passionately about ambitious action by all countries in order to hold temperature rise to a minimum. It knew that the goals it was devoted to could not be met if major developing countries were given a pass on taking robust climate action. At the April 2012 MEF meeting we hosted with Italy in Rome, Colombia's fiery and fearless negotiator Andrea Guerrero, fresh from the important role she played in Durban, took the floor with an intervention aimed directly at the firewall countries after an extended back and forth on the issue of equity: "We see a difference of thinking here about . . . what constrains development. For some this constraint is in mitigation commitments, but for us the constraint is others *not* committing to reductions. . . . At this moment we're facing irreparable damage to our countries. It is not time to discuss who's to blame and who should act. If we don't act now we won't be able to develop at all. Commitments need to come from all countries, in a gradient, that should be flexible and evolve over time, taking into account past circumstances and current capacities." *Earth Negotiations Bulletin* talked of "a discernible chasm" between these two factions and described LMDC countries as demanding fealty to the provisions of the 1992 Framework Convention at a time when that treaty "reflects a reality light years away from the 2012 global landscape," never mind the landscape of a new agreement in the 2020s and beyond.[2]

The United States and China

Having been hosted by Minister Xie in his hometown of Tianjin in 2010, I invited him in September 2012 to meet me for a two-day visit in mine, Chicago. The visit was the brainchild of Taiya Smith, formerly senior China adviser to Treasury Secretary Hank Paulson, who had recently joined my team. Taiya also suggested that we expand the format beyond our usual bilateral discussions to include "track two" teams of outside experts—something that Xie appreciated and that contributed to the sense of a deeper, more engaged relationship.[3]

To start things off, I took Xie and a few of his colleagues to see the Chicago Cubs play the St. Louis Cardinals at Wrigley Field, one of the two great vintage ball parks left in the major leagues.[4] Xie had never been to a baseball game and loved it. He was especially taken by a mistake made by the Cubs' third base coach, who held up a runner at the wrong time, stifling a rally. At our bilat the next day, Xie invoked the coach's error and said that the two of us had to avoid making mistakes like that.

I took him the next day to city hall to meet Chicago mayor Rahm Emanuel, an old Clinton White House friend, who showed Xie a good time. During the meeting in Rahm's office, Rahm was particularly interested in Tianjin's use of barges to clean up its Hai River since one of Rahm's causes was to spruce up the much-maligned Chicago River and its walkways. Rahm said he'd love to get one of those barges, and Xie just smiled. For the last part of the visit, Rahm took us outside for a stroll along the Chicago River to show off what he had accomplished so far. As the visit came to an end, Xie thanked Rahm for being such a gracious host and said he hoped they could be friends. Rahm, never less than himself, answered, "We're already friends. Where's my barge?" That night, I hosted all the US and China participants at one of Rahm's favorite steakhouses, Chicago Cut.

At our long bilat on Monday, Xie, in good spirits, started off expounding on the value of our relationship, saying that it was "cordial and in-depth," and that we could be frank with each other and show each other our bottom lines. This reminded him of the short pull aside we had in 2009 in Rome, the day before the MEF leaders meeting when he told me China couldn't accept reference to a 50 percent global cut in emissions by 2050. "We talked to each other about mutual understandings. . . . I think it helped us."

Perhaps because sports were in the air after the Cubs game or because sports are always in the air for me, I picked up on his praise of our relationship and told him a story about relationships involving two of my basketball heroes, Larry Bird and Magic Johnson—without presuming to elevate either of us to their Olympian heights. I explained how they were the two best players in the NBA all through the 1980s, and went at each other hammer and tongs, game after game, year after year, to see which team would win the NBA championship, and how because of that experience, not in spite of it, they became good friends.

We also spent time talking about our perennial topic, differentiation, which Xie acknowledged as our most difficult issue. We didn't break new ground. Xie made a comment that he probably intended to be constructive but I had my doubts. I told him that I should probably be pleased with what he'd said, but I wasn't so sure and wanted to tell him a story to illustrate. I said that my wife and I took a trip to Morocco back when we both worked in the Clinton White House and before we were married. We found ourselves in a dimly lit, musty old antiques store with beautiful vases and bowls, and even small, ancient painted doors. After Jen and I decided what we liked, I did my best to bargain with the owner of the shop because I knew that's what you were supposed to do in Morocco. After we had done the deal, he said, "And for you, pick out something else, whatever you like—it's free." I said to Xie, "I hope I'm doing a better job than that now!" He laughed his happy, red-faced laugh and then told me his own version of that story. When he and his wife married, they were poor teachers and used to go to a store after work to try to bargain for vegetables, and one day the owner said, "Here's the vegetables and you can take the pot too." So Xie, said to me, "I've made the same kind of mistakes." But then he added, "At least what you bought were antiques; I bought a pile of vegetables. That's what I am talking about—common but differentiated responsibilities."

* * *

I got my first real taste of the Kerry modus operandi on April 2, 2013. I was in Chicago, and he found me at my hotel to talk about an upcoming trip to China. He wanted to do a strong joint statement—both to get the Chinese to publicly acknowledge the seriousness of the climate risk and to establish a new high-level climate change joint working group to advance concrete technology and policy cooperation. The timing to launch the working group, a couple months before the annual Strategic and Economic Dialogue (S&ED), was perfect because it set the working group up each year to feed new announcements and initiatives into the S&ED, the highest-level annual US-China gathering, led by both presidents and a wide range of cabinet secretaries.[5]

As it happened, Minister Xie was coming to town a few days after Kerry's call for a MEF meeting, so we used his time in Washington to negotiate the statement, which was released April 13 during Kerry's visit. Xie and I were named heads of the working group, reporting to the leaders of the S&ED's strategic and economic tracks—Secretary Kerry and Treasury secretary Jack Lew for the United States, and State Councilor Yang Jiechi and Vice Premier Wang Yang for China. Over the next three months we identified and started initial work in five areas of cooperation, including heavy-duty vehicles, electric power systems, and energy efficiency in buildings and industry.[6] A few weeks after the joint statement was announced, Xie told me he thought it had a "big influence" both within China and internationally.

On April 9, 2013, during the days Xie and I were negotiating the specifics of the joint statement, he came to my home for dinner, following another good suggestion from Taiya Smith. Accompanied only by his translator, Xie was delighted by the whole evening. Mike Froman, who lived nearby, started things off on a jovial note by making a short surprise visit, complete with a bottle of Chinese Maotai, recalling the night in Shanghai in 2010 when Xie and my intrepid then special assistant Clare Sierawski traded shots of that wicked brew as I failed my drinking test. My wife, Jen, made a high-class, all-American dinner of steak, potatoes, and green beans. Ben, our eight-year-old, said hello, happily received the little present Xie brought, and went to bed. Jacob and Zachary, sixteen and thirteen, respectively, joined the dinner. Xie wanted to talk about US politics and was particularly interested in Jen's former service as a senior adviser to Hillary Clinton in the White House as well as at State. He was already sizing up Clinton as a potential next president. And he talked for some time about what life was like in the remarkable periods of change he had lived through in China, including the Great Leap Forward and Cultural Revolution. It was a special night, and neither of us ever forgot it.

* * *

On May 21, 2013, the president's chief of staff, Denis McDonough, chaired an interagency meeting to prepare for the president's June climate speech at Georgetown. Picking up on an idea from John Podesta, then running the Center for American Progress, which he founded in 2003, McDonough asked me to try to negotiate a joint statement between President Obama and China's new president, Xi Jinping, phasing down emissions of HFCs, a highly polluting, industrial greenhouse gas. The idea was to have the statement ready for the first meeting of the two presidents, scheduled for June 7–8 at the Sunnylands Center in southern California. Time was short, but as it happened, I was already planning to go to China in ten days, so we had a chance.

HFCs, used for cooling and refrigeration, were an intriguing target, subject to two different treaty regimes: the UNFCCC, as a greenhouse gas, and the 1987 Montreal Protocol, as a substitute for a gas that damages the ozone layer. While there was no chance to phase down HFCs within the UNFCCC given its politicized nature and quasi-consensus practice of decision-making, the Montreal Protocol presented a real opportunity.

It was built to phase out industrial gases, had expertise, was mostly free of the ideological baggage of the climate negotiations, had a Multilateral Fund to help countries implement its requirements, and stood as the single most successful environmental treaty ever, having actually saved the ozone layer. My colleague at State, Dan Reifsnyder, had been working on a potential phase down of HFCs using Montreal since 2009, and together with a handful of diplomatic allies, had secured the support of over a hundred countries. But given opposition from China, India, Saudi Arabia, and others, the idea of approaching HFCs at the presidential level made perfect sense.

When I raised HFCs with Xie on May 31 in Beijing, he was resistant, which was not surprising. China's chemical and appliance industries both profited from their use of HFCs.[7] Chinese allies such as India and Saudi Arabia also opposed a phase down. And neither the Chinese nor their friends would be happy moving a proposed regulatory initiative outside the friendly common but differentiated confines of the UNFCCC. He claimed, incorrectly in the view of our State Department lawyers, that since HFCs are a greenhouse gas, the UNFCCC would have to give its approval before the Montreal Protocol could get involved. Noting that the European Union was planning to put forward a proposal at COP 19 in Warsaw seeking UNFCCC support for the Montreal Protocol to take action—an unhelpful move since it had no chance to succeed—Xie suggested that the United States and China support the EU approach and "at least . . . show our positive attitudes on addressing the HFC issue." But we wanted to get the phase down done, not just show that we were the good guys. We left Beijing on June 1 without much progress, but agreed to resume our discussions by phone. The Sunnylands meeting was set to begin in six days.

Back in Washington, Sue and I negotiated with Xie and his team by phone every day, either early morning for us and evening for them or vice versa. Sue, who lives near me, often came to my house, and we'd crowd around a phone in my home office. Xie held fast to his incorrect claim that UNFCCC authorization was required. Over and over we tried to square the circle with language recognizing that HFCs would continue to be greenhouse gases accounted for under the UNFCCC, even if the Montreal Protocol were deployed to phase them down. As time grew short, I played our ace in the hole, saying if we couldn't resolve our language difficulties, President Obama would raise the matter directly with President Xi. As I had learned back at the Mexican MEF in Cuernavaca in 2010, the idea of letting leaders negotiate language is like kryptonite for the Chinese. A day before the Sunnylands meetings began, we landed on a carefully sculpted text that read in full,

> Regarding HFCs, the United States and China agreed to work together and with other countries through multilateral approaches that include using the expertise and institutions of the Montreal Protocol to phase down the production and consumption of HFCs, while continuing to include HFCs within the scope of the UNFCCC and its Kyoto Protocol provisions for accounting and reporting of emissions.

This slim parcel of sixty words became the singular concrete achievement of the Sunnylands meeting. And the deal proved consequential, both for setting in motion a process that led to the "Kigali Amendment" to the Montreal Protocol three years later and as a demonstration that the United States and China could make progress on climate change at the highest level.[8] Together with the Climate Change Working Group that Secretary Kerry had delivered in April, the Sunnylands deal started to establish climate change as the issue on which the two countries could best make real progress. The idea that I had raised with Xie at our first meeting in 2009—making climate change a positive pillar in our bilateral relationship—was beginning to bear fruit.

There was an odd little postscript to this story a few weeks later, which in the end, served mostly to burnish Sue's reputation as the Zen master of minimalism. Somehow, the Chinese had understood the final six words of the agreed statement to say, "*on* accounting and reporting of emissions," rather than "*for*." Something had obviously been lost in translation as both sides scrambled to get the thing done the day before the Sunnylands meeting. The choice of preposition made no real difference, but we had already released the official English version, so did not want to change it, and the Chinese felt the same way about the Chinese version they had released. Since the S&ED was happening in Washington on July 10–11, at which time both sides would want to refer to the agreement, we needed to fix this little snafu. To this end, Sue and I met with Xie and his team on July 9 in a dimly lit ballroom at the Wardman Hotel in Washington. We went back and forth for a while with no progress until Sue had an inspiration that the whole thing could be solved if quotation marks were put around the US version of the text when it was referred to in S&ED documents. The

Chinese could read the quotation marks as meaning that this was the text that the United States released (accurate or not), and we would be comfortable that it was in fact the text we released. The Chinese puzzled over Sue's proposal for a minute or two, pronounced themselves satisfied, and life went on.

Shaping the Debate for Warsaw

In late September 2012, in New York at the end of the UN General Assembly week, Sue and I hosted a brainstorming dinner at the Petrossian restaurant to talk about the contours of a future agreement with Vivian Balakrishnan and Burhan Gafoor (Singapore), Tim Groser and Jo Tyndall (New Zealand), Connie Hedegaard and Artur Runge-Metzger (European Union), and Luis Alfonso de Alba (Mexico). I suggested that the new agreement should be based on "nationally determined" mitigation pledges. Connie had reservations, worried that a nationally determined structure could not ensure deep enough emission reductions, but no one, including Connie, could articulate a politically viable alternative in the new "applicable to all" world of the Durban Platform. Still, Connie's concern was legitimate, so I suggested that we try to develop provisions to stimulate the strongest possible mitigation measures.

To this end, at our MEF meeting in April 2013, I proposed that we should build in a "consultative period" by requiring countries to put forward their proposed mitigation commitments six to nine months before a COP so the proposals could be scrutinized and critiqued by other countries, nongovernmental organizations (NGOs), analysts, and the press in order to goad countries into putting their best foot forward. The European Union liked this idea, but took it further at our September 2013 MEF meeting in New York. Connie argued that countries need to know that their proposed commitments will be assessed by the international community to determine whether they are ambitious and on track to hold the global temperature increase to 2°C, and suggested further that criteria or indicators be developed for such assessments. These ideas made sense in principle, but I thought we needed a lighter touch of the kind reflected in our "sunlight" proposal. I felt sure developing countries would rebel against the notion of being judged and attack a process they saw as incompatible with nationally determined pledges. And that's eventually what happened.

* * *

At a July MEF meeting in the beautiful old Polish city of Kraków, Tim Groser suggested that the new agreement have a mixed legal form, creating a legally binding obligation to submit mitigation commitments, but where the commitments themselves would not be binding at the international level. He proposed that the commitment be largely based on measures that were legally binding at the domestic level. He implied, though didn't quite say, that the agreement he envisioned would include rules, such as on transparency, that would be legally binding internationally. He defended this hybrid proposal as necessary to ensure broad participation in a new agreement. Although we didn't gin up Tim's proposal, he clearly made it with US needs in mind.

* * *

The most memorable part of the Kraków meeting involved an extraordinary coincidence. The meeting was held on Ulica Krupnicza street, which turned out to be just down the block from the house where Sue's mother grew up before the Nazis came. Still more amazingly, her mother had been saved by Oskar Schindler—one of the fortunate souls on Schindler's list. And to make all this even more moving, Sue's mother came with her to Kraków for our trip, visited the Oskar Schindler Museum in the old Schindler Enamel Factory, and found her own photo there from when she was a young girl.

The Warsaw COP

COP 19 began on November 11 in Warsaw in the indoor part of Poland's National Soccer Stadium, at a time of year when darkness fell each day in the midafternoon. I arrived on Saturday, November 16, and it was clear right away that developing country anxiety about preserving the firewall was still high. I came ready with a new line in my pocket: that for an agreement meant to take affect nearly thirty years after the Framework Convention created Annex 1 and non-Annex 1, and meant to last for many decades, the original categories could not be both operational—determining who was supposed to do what—and immutable. If operational, we would need to create a process for countries to "graduate" from one category to the other, as happens in various other international regimes.[9] If immutable, then the categories would have to sit on the shelf without operational consequence. But on Sunday morning, I had breakfast with Brazil's new climate lead, Antonio Carvalho, who was not ready to play my game. In the tradition of Itamaraty, Brazil's savvy foreign ministry, Carvalho was shrewd, capable, charming, and well-versed, though harder edged than Luiz Figueiredo or Andre do Lago. At breakfast, he told me "hell will freeze over" before Brazil lets go of the 1992 annexes in their full operational form.

* * *

The Warsaw COP made important progress, particularly on mitigation. Parties formally agreed on the crucial structural point that country mitigation pledges should be nationally determined rather than negotiated or dictated, and they agreed on several additional mitigation points. First, they settled on what to call the nationally determined undertakings—in the end, nationally determined *contributions* (NDCs) because for many countries, including the European Union, the word "commitment" is understood to mean legally binding. Second, they decided that an additional clause would be added (a fix proposed by Sue), saying that the use of the phrase "nationally determined *contributions*" would be *without prejudice* to the ultimate legal form of the Paris Agreement since those who wanted the submissions to be legally binding didn't want the word "contribution" to be read as implying the opposite. Third, they decided, as the United States had proposed, that countries should be expected to submit their contributions early, opening them to analysis and critique. Fourth, they decided that such early submissions should be called—*intended* NDCs rather than *initial*, since some thought "initial" implied that countries would be expected to submit stronger final targets rather than just having that option. Finally, they determined when early submissions should be made—the first quarter of 2015 (the Paris year) for those countries *ready* to do so rather than *in a position* to do so, because "ready" signifies that the submitting party is making its own subjective decision, while "in a position" suggests that others might expect the given parties to submit early because they objectively appeared to be "in a position" to do so. These all seem like minor points, but each one needed to be wrestled to the mat. And all of these decisions were sorted out not in a conventional meeting with twenty or thirty country negotiators sitting around a table but instead in impromptu huddles that took shape on the spot when more formal discussions appeared to get stuck, negotiators crowded together, standing up, with someone of broad credibility spontaneously taking on the role of moderator—often, in Warsaw, Singapore's trusted lead negotiator Burhan Gafoor.[10]

Interestingly, the consequential nonissue in Warsaw was that all countries agreed to the same words to describe their mitigation undertakings. Since the Bali Action Plan, proponents of the firewall had always demanded that different words be used to describe what developed and developing countries agreed to do on mitigation. Bali was regularly invoked to say that developing countries would agree to "actions," while developed countries would agree to "commitments."[11] In Copenhagen, we and developed country allies rejected the actions/commitments dichotomy, settling for targets versus actions, since "target" was at least a less loaded word than "commitments." But from Warsaw forward, the term of art for all became NDCs. The shift to a common term for country mitigation pledges was itself a significant move away from the firewall, as was the way it was accepted in Warsaw, scarcely rippling the water.

* * *

"Loss and damage" refers to climate damage resulting from either extreme weather events or "slow-onset" processes, such as the inexorable rise of sea level, threatening low-lying islands and coasts. It became an issue of mounting importance by 2012, especially for vulnerable countries on the front lines of risk. Toward the end of October 2013, as we prepared for COP 19, I had useful conversations with Marshall Islands minister Tony de Brum, who became a crucial leader on the road to Paris, and Peter Thompson, Fiji's permanent representative to the United Nations. Both were strong proponents of action on loss and damage, but also moderate in their approach and eager to work with the United States. I wanted to underscore that the United States was committed to strong climate action, we thought building resilience to reduce risk had to be the first objective for vulnerable countries, and we could work with them in designing a new loss and damage "mechanism," as we had all agreed to do at the Doha COP in 2012. I pointed out that over the past decade, the United States had been the world's leading provider of disaster assistance, but that we could not support demands for liability or compensation. The Warsaw International Mechanism for Loss and Damage was adopted at COP 19–a next step, but not an end point.

Postmortem

The Warsaw conference started to put some shape on an eventual agreement. It was going to have a nationally determined core, but with elements meant to promote ambition, clarity, and transparency. It took a consequential step regarding the legal character of the agreement by choosing the phrase NDCs—without prejudice, but still. The all-important Durban Platform phrase "applicable to all" was not abandoned, reinterpreted, or otherwise desecrated. And the Warsaw International Mechanism for Loss and Damage was established. In short, the gears of the negotiation process were beginning to engage.

The other notable positive feature of COP 19 was seeing the reputation of the United States clearly on the rise, owing importantly to the aggressive Climate Action Plan that President Obama laid out at Georgetown and was now implementing, and our stronger diplomatic engagement, featuring both the president's climate agreement on HFCs with Chinese president Xi and Secretary Kerry's high-profile climate activism. Minister de Brum from the Marshall Islands, UK minister of energy and climate change Ed Davey, and Tosi Mpanu, a prominent African negotiator from the Democratic Republic of the Congo all praised the climate action and engagement of the United States.[12]

But Warsaw also made it clear that the road ahead was going to be difficult. Elliot Diringer of the Center for Climate and Energy Solutions noted that Warsaw had managed "a modest package of decisions that keep the international climate negotiations on track but underscore the formidable challenges facing parties as they work toward a new global agreement in Paris in 2015."[13] Fiona Harvey, the astute climate correspondent for the *Guardian*, wrote in her wrap-up piece that "arguments over either keeping or redrawing [the] 'firewall' between developed and developing countries are likely to dominate the negotiations in the run-up to Paris." I was quoted as saying, "This is now the major faultline at the talks, and [the insistence by some] on deciding who does what in a new agreement based on unchanging 1992 categories is more pronounced than at Durban and poses the biggest challenge to the negotiations over the next two years." For good measure, Harvey also quoted Indian minister Jayanthi Natarajan in the same piece, underscoring that "the firewall exists and it will continue to exist."[14]

In a note of appreciation I sent to my superlative team the day after the COP ended, I said, “It was a tough two weeks and these negotiations will only get more challenging, I suspect, as we get closer to the D-Day moment when developing countries are supposed to join a legal agreement of some kind that applies, for real, to them. The reality of that Durban commitment, I'm quite sure, is the thing provoking this ‘double down’ on CBDR/annexes that we've been seeing all year and saw in spades in Warsaw. If one of our objectives was to fight back against this, expose the fundamental fault line of these negotiations, and make firmly clear what is and is not acceptable, educating the press about this at the same time, then I think we succeeded.”

7 The China Card

In late January 2014, Secretary Kerry called me up to the beautiful old sitting room outside his office, where he liked to tend the fire and tend to his yellow Labrador, Ben. We had made good progress with China in 2013 with the creation of the new Climate Change Working Group and the HFC deal between Presidents Obama and Xi. Now he wanted to do a new thing. Something big, something to keep up the momentum, something to accelerate it. This kind of restless attitude toward the world, toward seizing the moment and using the power of his office to break through, was always one of Kerry's greatest strengths. He told me to come up with an idea that he meant to take with him on his trip to Beijing in February.

That same day, I called our team together in the large, windowless conference room where we spent an inordinate amount of time. We knew that Obama and Xi would be together in Beijing in November for the Asia-Pacific Economic Cooperation meeting, so that would likely be the moment to announce a new initiative. After we discarded one or two other ideas, Clare Sierawski, my smart, irrepressibly positive, get-it-done aide who was variously my special assistant, China adviser, and chief of staff, spoke up. She suggested that Obama and Xi make a joint statement in Beijing announcing US and Chinese emission targets for Paris. Clare was across the table and two or three seats down from me, and I remember staring at her for an extra beat because the suggestion was so different from the kind of thing I was expecting. My first reaction was that it wouldn't work because the Chinese would be loath to get crosswise with their traditional partners from BASIC or the LMDC group. But a minute or two later it struck me that China's top leadership might see a significant upside for the overall US-China relationship that could overcome any hesitancy on the climate front. Xi's mantra in those days was "a new model of major power relations" by which he meant a harmonious approach between the United States and China rather than the United States trying to prevent China's rise, as would be predicted by the so-called Thucydides's Trap, where an incumbent great power and a rising power eventually come to blows.[1] I also realized that if we could make Clare's idea happen, it would be a game changer. Once President Xi walked down the aisle with President Obama, China would *need* to get a Paris deal done. Being satisfied to avoid blame in the event of a COP failure—the old playbook we had seen back in Copenhagen—would no longer work.

At the same time, it would be a high-risk gambit. We could only go through with the announcement in the end if China's targets were seen by the world as strong enough since it would reflect badly on Obama if he effectively blessed weak Chinese targets in a high-profile event with Xi. And we would likely not know what the proposed Chinese targets were until the fall. If we ultimately saw those targets as inadequate and had to scratch the big event, a plan with great potential to spur global climate action, our bilateral relationship could turn sour, with bruised feelings all around. On balance, I thought it was a risk worth taking.

I went back up to see Secretary Kerry the next day and told him about our idea. Like me, he wasn't enamored immediately, but soon warmed to it, seeing its potential benefits. He asked that we prepare a public statement about the idea that he could issue with the Chinese during his February visit, but I told him that wouldn't work because the idea of a joint announcement needed to be kept secret in case we had to pull the plug down the line. In addition, if we announced our plan for a joint statement in February it would ignite a lobbying and public relations frenzy aimed at pushing both countries to adopt targets that would almost surely be unrealistic, undermining the positive impact of our plan. Finally, I thought the chances of the Chinese leadership agreeing to box themselves in with a public announcement of our plans now were virtually nonexistent. All we needed now was for the Chinese to agree to put this effort in motion with the undisclosed aim of delivering a joint announcement in November. But Kerry was insistent on some kind of public statement for his February trip so that the Chinese couldn't walk away. And of course, no politician likes to launch a bold new effort without leaving his personal stamp on it. So I proposed a middle ground: a public statement during Kerry's February trip that the two sides would collaborate with each other on the development of our respective targets for Paris, together with a secret understanding of our aim to have the two presidents jointly announce those targets in November. Kerry agreed. I then sent a memo to John Podesta explaining our idea to make sure that the White House was on board. John said yes, and we were ready to go.

* * *

Kerry and I left for China on February 12, making an initial stop in Korea. While he met with his Korean hosts, I called Xie from our hotel on the evening of February 13. I walked him conceptually through what we had in mind, explaining that we would want to announce the idea of working together publicly during this visit, but keep private the aim of a joint announcement by our leaders later in the year. I told him both Secretary Kerry and President Obama were quite interested in the idea, and suggested that we could use the new US-China Climate Change Working Group as the vehicle for our collaboration, with a focus on sharing policy ideas and modeling results, as well as various assumptions we were each relying on in developing our post-2020 targets. I underscored that both sides would have to be comfortable with what the other was proposing before going ahead in the end since each side would be implicitly endorsing the other's target once our leaders locked arms. Xie seemed quite open to our thinking and said that China hoped our bilateral cooperation on climate change could become a "very bright spot in building a new model of great power relations." That Xie was already echoing Xi's mantra suggested that the Chinese saw the larger benefits of moving forward.

Xie agreed that the idea of a new cooperative effort between China and the United States would be made public, without reference to a possible presidential announcement, and that both sides would have to be comfortable in order for a joint announcement to go forward. But when we got to the matter of the public statement to be issued during our current visit, things got stickier. The Chinese had marked up a draft we had sent them, and it read as though it was all about continuing our existing efforts in the new Climate Change Working Group rather than about a new, enhanced collaboration aimed at Paris targets. We agreed that I'd send him a new draft and we would talk again in the early afternoon the next day in Beijing, though still by phone given the challenge of our schedules.

Early the next morning, Kerry and I flew to Beijing, and I accompanied him to his first meeting, with President Xi. Climate change was the first issue Kerry raised, and he discussed it at length, stressing the importance of US-China climate cooperation and the potential for climate to be an illustration of the new model of great power relations. He also talked about our idea of a joint presidential announcement of our respective targets, together with collaborative, preparatory work by the two sides. He told Xi that President Obama was interested in this approach and keen to elevate US-China bilateral efforts toward reaching a strong Paris Agreement in 2015. Finally, Kerry told Xi that he hoped to be able to announce on this visit the agreement by the two sides to share information and ideas related to developing our 2015 targets. Xi indicated that he supported the general thrust of what Kerry was proposing.

After the Xi meeting, I returned to the Marriott hotel to call Minister Xie about the proposed joint statement Kerry wanted to release the next day. There were two problems. First, Xie insisted on including the loaded phrase "under the Convention." I saw this as poison because following the Durban Platform, which, recall, included no reference to CBDR, the cohort favoring a firewall started arguing that "under the Convention" meant that a new agreement had to follow every element of the Framework Convention, every principle and provision; in effect to mean more than CBDR ever had. This was ridiculous, but it meant to me that I could not allow the phrase in our press statement. Xie responded that since the phrase was included in the Durban and Warsaw outcomes, developing countries would be quite unhappy if our statement left it out. I said we were now in a completely different, bilateral context and should keep the joint statement neutral in regard to contested negotiation issues. But he was dug in. After a lot of back and forth, I said I thought that for now, we had hit each other's bottom line—he had to have "under the Convention," and I could not accept it—so I proposed that we come back to the issue in an in-person meeting that night.

The other problem, less confounding but real, was that Xie opposed any clarity in the statement indicating that the United States and China would collaborate on developing targets for the Paris Agreement. He preferred much vaguer language about Climate Change Working Group initiatives. But indicating that the two sides would collaborate in working out their Paris targets was the whole point of the statement. To have any zing, the statement needed to say something interesting. US-China collaboration on Paris targets *did*; collaborating generally did not. I made a suggestion that retained a lot of the Chinese language but referred to "the sharing of information in developing their respective post-2020 plans to limit greenhouse gases." Xie wanted to delete any reference to the "development" of those plans, saying the word was too "specific." He wanted broader language. We left this point open as well.

I then rejoined Secretary Kerry for a short meeting with China's lead foreign affairs official, State Councilor Yang Jiechi, and a long one with Premier Li Keqiang. Kerry spent substantial time on climate in each meeting, and in his meeting with Li—a meeting Xie attended—he pushed our idea for a joint presidential announcement as well as a strong statement now about US-China collaboration. I then skipped the official dinner with Yang to regroup with Xie, Su Wei, and his team at around 8:30 p.m. in the usual conference room where we met at the NDRC. Uncharacteristically, and because I was traveling with Kerry rather than on my own, I didn't have anyone on my team with me, but was joined by Erica Thomas, an excellent foreign service officer in Beijing.[2] It took two hours to get this small piece of business done. Xie started off repeating his insistence on including the phrase "under the Convention," but suggested that we add the phrase "applicable to all" to address US concerns. But I had no give in my position. I said that "under the Convention" had been given a meaning by some countries, including China, that we could not accept and that UNFCCC phrases like these had no place in a bilateral agreement. I said if he had to have "under the Convention," we should all go home because I wouldn't accept it. Xie suggested adding some additional phrases referring to an "ambitious 2015 agreement" that would have "broad participation." This was his "spoonful-of-sugar-makes-the-medicine-go-down" strategy. I said no and added that if we could not agree to a joint statement, then so be it.

I then shifted to the other open issue in an effort to produce some kind of forward momentum in our discussion because of course I did want to get a joint statement done—not at any cost, but still, I had an eager secretary of state waiting for the thing and did not want to disappoint. So I suggested deleting the somewhat oblique reference in the draft statement to a 2015 agreement. Xie liked that, but wanted to delete any reference to the year 2015 altogether. I said he could either take out a reference to 2015 but give me the reference I wanted to "the development of" our post-2020 plans to limit greenhouse gas emissions, or vice versa, but he couldn't have both. He chose to give me the reference to 2015—which I thought was better for us. With that forward momentum—we had now accomplished something—he also stopped fighting about "under the Convention" because he knew I would walk away rather than accept it, and he didn't want that either. Finally, he pushed for a while to say "control" rather than "limit" emissions, but I borrowed a gesture from his own bag of tricks, acting incredulous that he could even suggest such a thing, and in the end he went along, at which point we were done. All of this, mind you, two long telephone calls and a long evening negotiating session, for a small statement of modest impact. The Chinese—talented, precise, clever, stubborn, self-confident, and endlessly patient—never made it easy. It was no wonder that progress came slowly.

The last hiccup in the statement saga came the following morning at the Foton-Cummins heavy-duty engine plant, where Secretary Kerry was scheduled to take a tour and make remarks. Kerry wanted to reference the new statement, but the Chinese had not run all the bureaucratic traps. To the distress of Kerry's State Department minders, who were obsessed with keeping him on schedule, I grabbed him as soon as he arrived at the plant and pulled him into a small holding room with Minister Xie, explaining the hang up. Kerry pressed Xie hard, leaning in on him like a Boston Brahman LBJ, ultimately saying, "Who do I have to call? Xi Jinping? Li Keqiang? Tell me who." Xie said he'd try to work it out. A few minutes later, Erica found me and told me to tell Kerry to slow walk the tour so Xie would have enough time to get clearances. I did, Kerry slowed down, Xie got the clearances, and Kerry talked about the new one-page agreement from a podium in the factory, with Xie and me next to him on the small stage.

* * *

In mid-March, President Obama sent a letter to President Xi that strongly endorsed the joint announcement idea, and Obama also raised it with Xi on the margins of a Nuclear Security Summit in Amsterdam, March 24–25. I saw Xie in March in Washington for a meeting of the Climate Change Working Group, and we talked then about arranging a set of meetings in Beijing on our new project, including technical teams from both sides, to be held a day or two before the upcoming 2014 S&ED.

* * *

The next bump in the road came May 9 in Mexico City, where Xie and I and Caroline Atkinson, who had recently taken over Mike's White House position, met on the margins of a MEF meeting.[3] For quite a while, it was an odd and uncharacteristic encounter. I led off, summarizing the basics of the joint presidential announcement we were aiming for, stressing, as I had before, that each side would have to be comfortable with what the other was proposing. This was true as a practical matter, but I underscored it as a way to pressure China into taking the strongest targets possible and talked a bit about the collaborative discussions we looked forward to our teams having in Beijing in July. But Xie wasn't himself, and seemed to be getting more and more uneasy as I spoke. He said that there had been "suggestions from [the US] side" that the two leaders would announce their post-2020 plans at the November Asia-Pacific Economic Cooperation summit, but that China wouldn't be able to finish the work needed to prepare its targets this year. Xie also expressed unease about the notion of each side "feeling comfortable" with the other side's target. He said he had only understood the notion of feeling comfortable to apply to the broad idea of doing a joint announcement.

I was completely taken aback and more than a little worried. There had of course been much more than "suggestions" from our side about a joint announcement; there had been agreement by both sides, all the way up to the presidential level—conditional in the end on both sides being comfortable—but still agreement. At our meeting in March, Xie indicated that China was working toward a November announcement and evinced no concern. In the meetings Secretary Kerry and I had done in February in Beijing, China appeared fully on board with the concept and raised no issue of timing. I had never experienced this kind of thing with Xie, whom I liked and respected, and in whom I had confidence. I tried to explain again the need for each side to be comfortable with the other's proposal, but Xie seemed to be getting more and more agitated, saying he had always believed there was mutual trust between us, implying that the conversation was giving him pause.

And then a few minutes later, I understood. He said, "I don't think *mutual consent* is necessary." And of course it was not. I knew that they would never accept *that* kind of formulation—China giving the United States a right to veto its target would indeed undermine Chinese sovereignty and vice versa. Something crucial had been lost in translation; "feeling comfortable" had been rendered as "mutual consent." Once I understood, I jumped in and said that there had been some confusion. Each side's target would be nationally determined and in no way subject to the consent of the other side. I gave him an example of what I meant by feeling comfortable. I said suppose we decided that our own target for 2025 should be an 18 percent cut below the 2005 level, only a 1 percent improvement beyond our 17 percent target for 2020. We would have every right to do that, I noted, but Xie might tell me that President Xi was unwilling to stand up next to President Obama as Obama made such a paltry announcement. China would have no right of consent, but every right to say it did not want to do the event. When I finished, Xie said, "You've made yourself very clear," and all the earlier agitation melted away. We were back on track.

I subsequently got wind of a misguided move afoot in the White House (not involving Podesta) to go over Xie's head and complain at a high level about his conduct in the Mexico meeting. First I was dumbfounded and then I was furious. Around thirty seconds later, I contacted the White House with an urgent message to shut down this amateur effort at once. This was a four-alarm fire for me. Lodging a White House complaint against Xie would have been entirely unfair to him and done enormous, probably permanent, damage to a relationship that both of us had painstakingly nurtured, and that was central to both countries and global efforts on climate change.

* * *

In early July, our interagency team traveled to Beijing for both technical and political meetings. John came as well. On July 8, the technical teams met with each other while John and I met with Xie and his negotiating team, after which the technical and political teams from each side gathered in a large, combined meeting. Xie took the floor early and launched into a dissertation on China's domestic climate efforts in mind-numbing detail. At first I thought he was just displaying the Chinese penchant for more technical detail than nontechnical people can bear, but as he went on and on I realized that this was an actual filibuster. He wanted to run out the clock and he did. We were initially worried that he might be trying to back away, but it later became clear that he just didn't want to talk about our private objectives in such a large gathering. In the late afternoon, John and I joined Xie and Su Wei at the Diaoyutai State Guesthouse for a two-on-two meeting that occurred in an odd space that was a kind of corridor with a high ceiling, some fifteen large chairs on each side, yards of space in between the two sides, and no table. But we got right down to business and it was clear that the game was still on. We also met the next day with

Vice Premier Zhang Gaoli at the beautiful Zhongnanhai Leadership Compound. It was a formal meeting, and Zhang is a formal man, rail thin and dour, with a ramrod back.[4] He wanted us to know that he thought the joint announcement idea was important and he had personally supported it.

* * *

In the early fall, internal discussions intensified about what our own Paris target would be. The analytic work to support this decision was led from the White House, and my only concern was that we not be too cautious. The White House team was working hard to make sure that when we put forward a number, we could see a clear path toward achieving it, and that was a good basic instinct. At the same time, I thought there were plausible factors that could help push the envelope, such as overcompliance by states with our new power plant regulations. Plus I thought if we were to put forward a target that was, say, 2 percent beyond our true confidence level and ten years later (2025) we came up a bit short, having made our best effort, that would not be a hanging offense. I thought better to get caught trying with a stretch goal than be too cautious, but I also didn't believe we would come up short. And pushing a bit beyond our safety zone would help prod other countries to do more as well as bolster our own leverage and credibility, putting us in a stronger position for the tough final year of the negotiations. Still, mindful of the instinct for caution among some, I ultimately suggested that we should express our target as a range so we could provide a little reassurance on the downside while still getting the benefit of a stretch goal on the upside. And this is what we did in the end.

Our team also analyzed a sizable volume of research on what targets would be strong enough on the Chinese side to justify Obama standing up with Xi. We assumed, as a starting point, that China would use the same kind of targets it had submitted after Copenhagen for the period up to 2020—especially a reduction in the carbon intensity of its economy and increase in the percentage of nonfossil primary energy it used. We were looking at both what we thought would be a strong enough step forward and how we thought informed observers and the press would react.

* * *

In early September, President Obama sent a second letter to President Xi. It focused on what could be accomplished between the two countries during the president's November visit to Beijing and again underscored Obama's keen interest in the joint announcement on climate change. Meanwhile, John and I were eager to start talking with the Chinese about the specifics of our respective targets, with actual numbers, even if still tentative. We hoped to do that in a meeting with Minister Xie and Zhang in New York on the margins of the annual UN General Assembly meetings, the week of September 22. During those same early September days, Xie started lobbying hard for a meeting between President Obama and Vice Premier Zhang during the UN General Assembly session. President Xi was not coming to the session and neither was Premier Li, so Zhang would be the highest-ranking Chinese official in New York, but it was still a real protocol stretch to ask Obama to meet with him. Yet it became a cause for the Chinese. Xie pressed us repeatedly to confirm the meeting, and avoided confirming our own session with him and Zhang until the Obama meeting was locked down. Cui Tiankai, Chinese ambassador to the United States, invited John and me to lunch at the embassy, emphasized how committed the Chinese were to the joint announcement, and then pressed us some more for an Obama-Zhang meeting in New York. The stakes were clear enough, so John got the meeting scheduled. When I called Xie to tell him the White House had accepted the meeting, he put the good news in his pocket and without missing a beat started right in on me about making sure there would be ample time for the meeting so they could have a good discussion—not just one of those short pull asides. To bolster the odds of securing this plussed-up request, he told me that Zhang would be carrying a message for Obama from President Xi. I rolled my eyes.

On September 23, Zhang got his meeting with Obama at the United Nations. Soon after that, to prepare for our meeting with Xie, John and I huddled with our team at one of those couch-chairs-coffee-table setups that dot the UN headquarters. The main question was whether to share our preliminary numbers with the Chinese even if they were not yet prepared to share theirs. Most of our group thought we should not. This was perfectly sensible, but I disagreed. I had no confidence that the Chinese would show us their numbers, even though we had already discussed sharing numbers with them, but I argued that we should share our initial numbers anyway because I wanted them to know how significant our numbers were if they were still making final decisions about their own, and I also wanted to eliminate any excuse for them not showing us their numbers soon. John didn't seem altogether convinced, but he backed me up. At the meeting, Xie indeed said they were not ready to share numbers, but I laid ours out, saying that we could reduce our emissions by 25 percent compared to 2005 and might go higher, but hadn't decided yet. We wanted to hold back where we might go on the high end so we had something in our pocket to use as talks progressed.

* * *

Several weeks later, on October 16, we had a video call with Xie and his team, again set up to discuss proposed numbers, and again Xie came with nothing but weak excuses about having to prepare for an upcoming Communist Party meeting. This behavior was in line with what Trevor Houser, my China hand from 2009 whom I had signed up to help on the joint announcement project, had expected. He had predicted from the beginning that the Chinese would try to lock down arrangements for the joint announcement itself, so the event was set in stone, while dodging any extended discussion of their numbers. Sure enough, after saying he had no numbers, Xie pressed to discuss the plans and logistics of the joint announcement. We ended the call.

* * *

On Saturday, October 18, Secretary Kerry asked John and me to join him and State Councillor Yang Jiechi for lunch at Legal Seafoods Harborside in Boston. Kerry was hosting Yang for an informal visit. They had dinner at Kerry's home the night before with Kerry's wife, Teresa, and then meetings that morning at the Taj Hotel across from the Public Garden (formerly the iconic Ritz Carlton). The lead China advisers at State and the National Security Council, Danny Russel and Evan Medeiros, respectively, attended, as did Kerry's deputy chief of staff, Jon Finer, while Yang was joined by Ambassador Cui Tiankai and Assistant Foreign Minister Zheng Zeguang. Kerry talked at length about climate change. Yang was cordial but noncommittal and not nearly as steeped in the issue as Kerry. But the key message Kerry delivered was that the joint announcement was still an "if" proposition, not a done deal. It remained conditional on the targets of both sides being ambitious enough that the international community would receive the announcement with enthusiasm.

* * *

Along with John and the team, I flew to Beijing on Sunday, October 26—less than three weeks until Obama's visit—to try to nail down the joint announcement. We met with Minister Xie and his team on Tuesday morning at the NDRC offices, and this time the Chinese got right down to business. Xie told us they had three targets affecting the full economy plus an afforestation target. He said China would announce one or two of the main targets now and the other(s) in 2015. The main targets were peaking CO_2 emissions "around" 2030, improving the carbon intensity of the economy (amount of carbon per unit of gross domestic product) 60–65 percent by 2030 compared to 2005 levels, and increasing the nonfossil share of primary energy to 20 percent by 2030. Of these, the peak-year target was the most distinctive since for the first time, China was proposing a reasonably hard target for stopping the growth of its economy-wide emissions. The time frame was longer than we wanted to see and we didn't like the hedge word "around." But it was still notable. The nonfossil target sounded to us like the toughest. Trevor calculated that China would have to add approximately eight hundred gigawatts of nonfossil energy by 2030 to meet the target, a rate of fifty gigawatts per year—the annual equivalent of approximately a hundred full-scale, five-hundred-megawatts power plants. The carbon intensity target was the least impressive because it seemed too close to a business-as-usual trajectory. John and Xie debated the point for some time without any progress.

After a short lunch break, John said, "Your carbon intensity numbers put the joint announcement in jeopardy. Let's focus on the peak year." I then said we thought 2030 would be relatively easy for them and that 2025 was quite feasible for a number of reasons, including Xi's broad plan of restructuring the Chinese economy away from so much reliance on heavy industry. Xie said we were not negotiating; the targets had already been cleared through senior levels of his government. He attacked our proposed 25 percent emissions cut, though I told him we were looking as high as 28 percent. He said to us, "You are worried about our numbers, but we are worried about yours!" Much of this was just sparring. I didn't think Xie was concerned about our numbers; he was just deploying his standard tactic of going on offense when criticized. For us, the Chinese numbers were vitally important. My initial reaction was that we could live with what they told us, but we continued nearly until the end to do analysis and assess how we thought their package would be received.

After lunch, we met with Xie and Zhang at the Great Hall of the People and covered a lot of the same ground. John told Zhang that each side argued that the other's wasn't strong enough, and Zhang replied that Xie had said the same thing. Zhang defended China's targets, but mostly was in the mode of focusing on the big picture, urging us to do the deal, saying that the joint announcement would be a "bright spot" in our bilateral relationship. He noted that "the United States proposed the joint announcement; accepting this was not easy." Zhang urged us not to be "so intense," and said "we must understand each other and be friends. The most vigorous negotiations are the most successful." He said that Xi and Premier Li were "resolute" in wanting the joint announcement to happen.

After the meeting with Zhang, we returned to our hotel and continued to discuss and debate. Some of the younger members of our team thought that Xie was bluffing when he said China couldn't change its numbers, but I didn't think so, and neither did John. I thought that whatever leverage came from China wanting to strengthen its targets to satisfy the United States was already baked into its numbers, that its targets had indeed been cleared at a senior level, and that the Chinese would not budge, especially given their regular refrain about sovereignty, deeply engrained in the national psyche, and repeated that day by Zhang: China was not acting "because others require us to do this, but because we must do this for us."

John and I, together with Sue Biniaz and Michelle Patron from our National Security Council staff, joined Xie and members of his team for dinner at the Diaoyutai Palace. As we were milling about before dinner, having hors d'oeuvres and drinks, we tested Xie one more time to see if we could change "around" 2030 to "by 2030," but he would not engage, reiterating that China's targets were final and not changing. We thought the Chinese targets probably cleared the hurdle of being strong enough, but wanted to talk it through. The whole exercise of trying to define the bottom line of what was acceptable was fraught, as much art as science. Analysts and modelers were all over the map on what the Chinese could or should do. And the abstract exercise we had engaged in before getting real numbers from China was inevitably different from staring the actual numbers in the face, especially given the historic significance for climate negotiations and the bilateral relationship of the two presidents making this announcement together.

After dinner, John and I talked to Obama's Chief of Staff Denis McDonough, National Security Adviser Susan Rice, and the president's science adviser, John Holdren, a longtime leading climate change expert, laying out the pros and cons of the Chinese numbers. The most important voice on the call was Holdren's. He supplied the crucial, well-informed opinion that the Chinese pledge to peak their emissions around 2030—the marquee part of their proposal—was OK. Not as strong as we would wish, but not bad. That opinion solidified John's and my view. And Denis and Susan, focused on the potential foreign policy power of the joint announcement by Obama and Xi, were more than satisfied.

Even after the call to the White House, though, I continued to worry about whether China's targets would be well received in the climate world. It wasn't that I personally thought the Chinese numbers were off. Quite like Holdren, I would have been happy with more, but especially in the wake of his advice, I thought its proposal was strong enough. Yet the stakes were high. If we sent the president out to make the joint announcement with Xi and the Chinese numbers were badly received, the whole effort would have failed, thereby undermining the climate negotiations and damaging the bilateral relationship. In short, we were up on the high wire without a net. Even with Holdren's advice, I wanted more reassurance. So at around 2:00 a.m., unable to sleep, I went down to the large, deserted lobby of the Westin Hotel to call three of my wisest, most trusted international colleagues—Bo Lidegaard (Denmark), Jo Tyndall (New Zealand), and Burhan Gafoor (Singapore). I asked each a strictly confidential question about what their reaction would be if, hypothetically, China were to announce a target of peaking its emissions in 2030 as part of a joint initiative with the United States. I did not explain the ongoing negotiation or the context of a potential presidential announcement. All three thought the number was certainly good enough to justify going forward with such an initiative even though hardly perfect.

John and I met again with Xie and his team the next morning, October 29, but despite the pep talk from Zhang and our call to the White House, things went downhill. Although we had been quite clear on Tuesday that the target we liked least was carbon intensity, Xie announced that the two targets President Xi would announce with President Obama would be carbon intensity and the peak year. That's when John got mad: "We have strived to move forward with different targets we find to be acceptable," John said, "but you are screwing around by proposing a target we don't like and rejecting a target we do like." Xie was not happy. After arguing with John for a minute or two about a different point, he said, "As for 'screwing around,' I never do this with my partners and friends. . . . How can you say screwing around?" John answered, "I want to get to yes, but I'm having a hard time seeing how. This is not productive." But in the end it was. China eventually let us know that it had decided to announce the peak year and nonfossil target at the joint announcement, holding carbon intensity back until 2015. John's left hook had landed.

We did a video call with Xie on November 3 from Washington, and John pressed him again to change the peak year from "around" 2030 to "by" 2030, telling him, as I had suggested, that Obama would raise it himself with Xi if need be. The threat to take the issue up to the top, which had been effective in our HFC negotiations with Xie in advance of the 2013 Sunnylands meeting, did not work here. Xie repeated that China's bottom line was peaking "around 2030." He also insisted that the text of the announcement include CBDR/RC, consistent with what he called the "multilateral consensus," and said that Zhang had told this to President Obama at the United Nations. I said that we had discussed this many times and understood each other's views about CBDR/RC very well, and suggested that we let the "two Sues" try to find language. Xie said that the announcement had to be acceptable to the world and that we had to show the world we were "conforming to Framework Convention principles." But he agreed that the next step should be a session between Su Wei and Sue B.

On Saturday, November 8, while I was on my way to Dulles airport to get my flight to Beijing with Sue and others, John called to suggest that we make a few additional calls—despite the risk of a leak—to get one last check that an announcement with a peak year around 2030 would be applauded. As I had done in the Westin lobby ten days earlier, John wanted to secure some last-minute reassurance. I called Selwin Hart, the Barbados diplomat and a good friend of mine then working on climate change for UN Secretary-general Ban Ki-moon, Martin Lidegaard, Denmark's climate minister (and Bo's brother), and France's Laurence Tubiana. John called the UNFCCC's Christiana Figueres. These soundings were all positive, and taken together with some good additional data from analysts regarding the ambition of a 2030 peak year for China, fortified us in feeling that we could live with around

2030, especially if we could shore it up with some good additional language in the text of the announcement. In my brief conversation with Laurence on November 9 as I paced outside a Beijing restaurant where I was meeting colleagues for dinner, she urged us to secure language in the announcement text emphasizing that our 2015 targets were part of a longer-range decarbonization effort—a point Christiana had also made to John.

* * *

Sue met Su Wei on Monday morning, November 10, to try to land language on CBDR/RC that would be acceptable to both sides. When the Chinese insisted that CBDR be included in the statement, we insisted that the language had to convey something about evolving or changing circumstances going forward, much as we had done during the Durban Platform negotiation. As she almost always did, Sue found a way, with new agreed-on language that made its way into the final text.

John arrived on November 10 with the president, and on Tuesday, John, Sue, and I sat down with Xie and his team for a final negotiating session on the language of the announcement. Xie agreed to add language that China would make "best efforts to peak early" and agreed to a version of what Laurence and Christiana had suggested about situating our 2015 targets within a longer-range decarbonization effort.

On Tuesday, Obama had a one-on-one tea with Xi where he planned to make one last run at changing "around" to "by" 2030. I had actually bumped into the president in the Westin fitness center that morning around 7:00 a.m. He was lifting weights and going through his workout routine with Secret Service agents nearby, while I was on a stationary bike, keeping the ritual I followed every morning on every trip I took around the world. White House staffers gave the president a wide berth, but he saw me, and after a while came over to say hi and ask whether we were making progress on the announcement text. I said, "Yes, a little, but it's painful." He said, "It's always painful. By or around." I said, "You're on your talking points." But at his tea, Xi wouldn't budge on "around." The Chinese had never been bluffing.

The other highlight of Tuesday was a meeting at 3:00 p.m. with Secretary Kerry and Zhang, which John and I also attended. Kerry spoke at length about climate, and Zhang went on for some time himself. The preparations at this point were done, but it was still useful for the two of them to talk.

The surprise came at the end, when the slightly awkward, rigid Zhang, who would never rank high on a list of likely huggers, stood up and hugged Kerry, John, and me in succession—a nice, if surprising, gesture.

In the end, the announcement text included several important features in addition to US and Chinese targets. In the core paragraph (¶3), the two countries set forth their 2015 mitigation targets in a precisely parallel sentence structure, consistent with our insistence on legal symmetry. As noted, China agreed to make best efforts to peak before 2030, while the United States agreed to make best efforts to hit the top range of its emissions reduction target of 26–28 percent below its 2005 level by 2025. In that same paragraph, the two countries agreed the targets they were announcing were "part of the longer-range effort to transition to low-carbon economies, mindful of the global temperature goal of 2°C"—the language inspired by Laurence and Christiana's suggestions. Paragraph 4 amounted to a powerful call to arms by China and the United States to get the Paris Agreement done. It said that the two countries "hope that by announcing their targets now"—four and half months earlier than the period the Warsaw and Lima outcomes had identified for early submission of targets—"they can inject momentum into the global climate negotiations and inspire other countries to join in coming forward with ambitious actions as soon as possible," preferably by the end of March. And it said that "the two presidents resolved to work closely together over the next year to address major impediments to reaching a successful global climate agreement in Paris." This was a powerful message—the leaders of the United States and China declaring for all the world to see that they intended to join forces to clear away any roadblocks and get the Paris Agreement done.

Paragraph 2 included the new CBDR language negotiated by the "two Sues." It said that the two sides were committed to reaching an ambitious 2015 agreement "that reflects the principle of common but differentiated responsibilities and respective capabilities *in light of different national circumstances.*" While we would have liked to say "evolving" national circumstances, the words Sue secured were still important. They said, in effect, that in this new agreement CBDR/RC would be based on circumstances—which are inherently dynamic and changing—not categories. And differences in circumstances manifest themselves across the broad spectrum of countries, including among developing countries, not just across the divide between developed and developing. This was the first change to the CBDR/RC catechism since 1992, situated in a groundbreaking document between the presidents of the two leading climate countries in the world. In addition, the sentence said that the Paris Agreement would "reflect" this new CBDR principle, which meant not that the agreement needed to be festooned with multiple references to CBDR/RC but rather needed to reflect this new kind of differentiation overall, which it might arguably do through devices such as a nationally determined structure.

Finally, the two countries announced a set of new joint efforts on clean energy research and development; HFCs; carbon capture, storage, and utilization; promoting trade in green goods; doing clean energy demonstration projects, and launching a new Climate Smart / Low-Carbon Cities Initiative.

The Joint Announcement

On Wednesday morning, November 12, at 7:00 a.m., John and I did a background briefing for the press, embargoed until after the actual announcement occurred, and then we were off to the Great Hall of the People, where we did a short, stand-up meeting with President Obama and others on his staff. Then Obama joined President Xi for the press conference in a large, high-ceiling hall in which they walked side by side down a long red carpet, took their positions behind two podiums, and then announced their climate change agreement. The reaction was immediate and astonished. The press, other countries, and close climate observers were caught completely off guard, which only amplified the impact. China had, at least for this important moment, stepped away from its traditional developing country allies and made common cause with the United States. A former colleague, then in the midst of a track two climate dialogue in New Delhi, emailed me in the moment to say that the Indians in the meeting were "shell-shocked." Somehow, although we had always known our secret effort might leak and had our press points ready for such a contingency, it never did.

The implications of the deal for climate change were momentous—the presidents of the two most important players, historic antagonists on opposite sides of the developed/developing country divide, had come together to announce their targets and pledged to work together to get Paris done. Suddenly countries all over the world that had every reason to be nervous about whether the deal could be struck in Paris, given the disappointing history of climate negotiations, sat up and started to believe that an agreement in Paris would get done. The *New York Times* said in its lead editorial the next day, "The deal jointly announced in Beijing by President Obama and China's president, Xi Jinping, to limit greenhouse gases well beyond their earlier pledges is both a major diplomatic breakthrough and—assuming both sides can carry out their promises—an enormously positive step in the uncertain battle against climate change. . . . The climate accord represents a startling turnaround after years of futile efforts to cooperate in a meaningful way on global warming."[5] And it was front-page news around the world.

The agreement was also resonant far beyond climate change. Sandy Berger, President Clinton's national security adviser, called the joint announcement the "most important agreement between the United States and China in more than twenty years." Max Fisher, then at *Vox*, wrote that the deal was "arguably as significant on pure foreign policy terms as it [is] on environmental terms. It sets a precedent of the United States and China not just cooperating on a difficult issue . . . but cooperating on global leadership."[6]

In the end, Clare's original idea changed the game, as on second thought, I realized it might. China was now hugely invested in getting a successful agreement done. It had staked the credibility of its president on it. There was no turning back. And that was the best news anyone could have delivered on the road to Paris. At the same time, as John and I got ready to do our press briefing that morning in a cramped room at the Westin Hotel, my thoughts had drifted to COP 20 in Lima, only nineteen days away, and what came to mind was that China would take as hard a line there as ever. The Chinese would want to reassure their surprised and potentially alarmed allies in BASIC and the LMDC group that they were still faithful to their shared principles and had not gone over to the other side. China *was* now fundamentally bought into the success of the Paris COP, but I expected that we would still be in for the usual rough ride when the curtain rose in Lima.

8 The Far Turn

COP 20 opened in Lima, Peru, on December 1, 2014, less than three weeks after the front-page, worldwide news of the US-China Joint Announcement. But while that news gave a potent boost to prospects for reaching an agreement in Paris in 2015, the mood in Lima was anxious. Lima was, after all, the last major station stop before Paris, so negotiators from all over were well aware that anything they agreed to at COP 20 would influence the outcome in Paris, and in turn, the nature of the international climate regime for a long time to come. They were also aware that if the Lima COP foundered—failing to make the decisions needed to guide countries in preparing their "intended nationally determined contributions" (INDCs) or even failing to produce any agreed outcome—it would threaten the mission to deliver a major climate agreement a year later in Paris. Two issues stood out as the most contentious among the many being negotiated: differentiation and whether INDCs would be subject to some form of assessment or review.

Differentiation

The most contested differentiation issue in Lima concerned an annex with options for the kinds of information countries would be expected to submit with their emission targets in order to "facilitate clarity, transparency, and understanding." The problem was that the annex included options that were bifurcated, with one set of requirements for developed countries and a less stringent, voluntary set for developing countries. The substantive differences were not earthshaking, but the principle mattered. If our side acquiesced in bifurcation here, it would establish an untenable precedent for the much bigger issue of how differentiation would be handled in Paris. For us, the divided options in the annex were a red line.

At a dinner that Sue, my deputy Trigg Talley, and I hosted on December 8 for Miguel-Arias Cañete, the European Union's new climate commissioner, and his number two, Jos Delbecke, Tim Groser and Jo Tyndall from New Zealand, and Laurence Tubiana from France, we talked about whether we should refuse an outcome if it included a bifurcated information annex. I said yes, sticking to my standard principle that if you believe you can't walk away from a negotiation, the other side will sense that and exploit your weakness. Some worried that blocking an outcome would kill Paris. Cañete said that the European Union would stand with the United States on this point. That was the start of a productive and genuine friendship between the two of us. I liked Cañete as soon as I met him. He is a husky guy with a halo of white hair and white beard; he speaks fast, staccato, Spanish-accented English, spitting out his words when he's amused or annoyed. He's sure of himself, has a sharp mind, a quick wit, the soul of a rascal, and a passion for race cars—which in 2014, he still raced. His stance here was a good sign, to my eye, that he was also tough and strategic.

The question of differentiation also arose in the context of how or whether to incorporate the classic CBDR principle in a Lima decision. The phrase was included in its original form in a November 11 draft text issued three weeks before the COP began. A December 8 draft text included it twice—once in its original form (as an option in a mitigation paragraph), and once in a new formulation that went beyond what we had done in the Joint Announcement, referring to "evolving" CBDR/RC.[1] Then the phrase was deleted altogether in the next drafts, on December 11 and 12, the last scheduled days of the conference.

Differentiation came up in two of my meetings with Minister Xie. On December 8, I asked him what he was thinking about differentiation, and he answered, "According to our Joint Announcement, we already solved" it. I thought it was a good thing that he assumed our Joint Announcement version of CBDR would become the generally accepted way of referring to it, though I did not think it had solved all differentiation problems. I assumed that China would still seek bifurcation between what developed and developing countries were supposed to do on matters including mitigation, transparency, and finance, despite our new Joint Announcement language. I said this to Xie, and he answered by reflecting, once again, on our pull-aside meeting in Rome back in 2009—an iconic moment for him—when I understood his red line: "Over the past years of cooperation,

I have found our communication very productive. We will continue to do so in Lima and Paris. . . . You must remember that as early as the meeting in Italy, we started consulting with each other on all major climate issues and we have managed to handle these issues." In effect, he saw the CBDR language in the Joint Announcement as emblematic of the way each of us tried to get what we needed while at least being mindful of what the other needed as well. There was certainly something to that, but it was also true that our side could only reach that place of mutual accommodation with the Chinese after Xie and Su Wei had done their level best to pick our pocket—though I suppose Xie and Su Wei might have said the same thing about me and Sue.

On Thursday, December 11, a day before the scheduled end of the conference, at a time when draft texts had included CBDR in different formats, Xie suggested that we support using the Joint Announcement formulation to resolve the issue, and I agreed. Xie told me that those last six words ("in light of different national circumstances") were "fixed in my brain," and I said same for me. Reflecting on the influence of the recent Xi-Obama accord, he said that he thought our Joint Announcement had embodied "the spirit of cooperation," and added that in all of his bilateral meetings, he heard praise and congratulations for the Joint Announcement: "many people are saying that if even the United States and China can reach agreement, there are no issues too difficult to solve."

As we were wrapping up Thursday evening, Xie suggested that we get together again the next day to see what sticking points remained so we could try to figure out how to resolve them. That comment in and of itself was notable—unlike anything he had ever said before the Joint Announcement. We now had, in effect, a kind of partnership—competitive, sparring, and contentious, but a partnership nonetheless, authored by our presidents.

Ex Ante Review

The other especially contentious issue in Lima involved the European Union's proposal to establish a process to review country INDCs—a proposal that came to be known as the "*ex ante* review" since it was meant to take place before submission of the final NDCs. When the United States proposed the idea of a consultative period at the May 2013 MEF meeting, we thought it would encourage countries to be ambitious so that they would look good in the eyes of the press, analysts, NGOs, and other countries. We avoided a structured approach because we assumed many developing countries would protest and that much of the salutary purpose of a review could be achieved without formality. In a more recent submission to the UNFCCC, the European Union had toned down the rigorous process Connie Hedegaard had called for at our September 2013 MEF meeting, describing a process that was "not . . . overly designed or prescriptive," intended to create "a common understanding."[2] But this could not erase the impression Connie had created.

An *ex ante* review was the issue that Minister Xie was most obsessed about in Lima, mostly because he thought that if there were an assessment process before Paris, countries would be expected to make their targets more stringent than the ones they had put forward, and he hated that idea. It ran counter to China's core political priority of avoiding interference with its ability to manage its own economy. If the EU plan went forward, Xie said to me on December 8, it "will will make Paris a disaster. . . . China will not accept an *ex ante* assessment in 2015."

At our meeting on Thursday, December 11, Xie had not moved an inch on *ex ante* review. I tried to reassure him that the European Union, at this point, did not intend to force countries to change their targets and urged him at least to be open to a process focused on clarification. But Xie did not believe the European Union had changed its views. He had heard Connie at the MEF in 2013 calling for a full-blown assessment and had seen a detailed process reflected in earlier versions of the Lima decision text. China's opposition to an *ex ante* review, however, did not win China friends among progressive and vulnerable developing countries. On Friday, December 12, Marshall Islands minister Tony de Brum, an influential voice for high ambition, declared from the floor of the plenary hall, "We are shocked that some of our colleagues would want to avoid a process to hold their proposed targets up to the light. If they are at the top end of their national potential, there should be nothing to hide." But China would not budge.

Looking at what successive draft texts said on the *ex ante* review process from November 11 through December 12 and then the final text on December 13 was like watching reverse time-lapse photography of a plant becoming smaller and smaller until it finally disappears into the ground. The November 11 and December 8 drafts each included over thirty lines in multiple paragraphs, laying out a detailed process. The December 11 draft included three different options: eliminating a review altogether; calling for "dialogue" among the parties in June 2015 with a technical paper prepared in advance; and calling for a more full-fledged review intended to assess the aggregate effect of the parties' INDCs as well as how each party's INDC constituted a fair and equitable contribution by that party. The December 12 draft had shrunk down to six apologetic lines calling for a "nonintrusive and facilitative dialogue" in June 2015, "respectful of national sovereignty," with the objective of facilitating clarity, transparency, and understanding of the INDCs, but only for "those parties willing to do so." Thin gruel. But Minister Xie wasn't done yet.

* * *

The Lima COP needed to resolve various other undecided issues as well, including what the main text—as distinguished from an annex—would say about the information parties needed to provide with their NDCs, since the annex was going to be dropped; the scope of what NDCs should cover—mitigation only or also adaptation and finance; whether wealthy developing countries (not just developed countries) should be encouraged to provide financial assistance to poor countries; and language making it clear that developed countries could not use the nationally determined character of NDCs as an excuse to "backslide" with new targets weaker than their old ones.

I had originally been skeptical that the backsliding concern was genuine since I couldn't imagine developed countries trying to reduce their ambition, but during a June visit to Brazil, Brazil's Antonio Carvalho persuaded me that developing countries were actually concerned about this. The persuasion took place, through an odd set of circumstances, in Antonio's spacious backyard in Brasilia during a dinner that included Luiz Figueiredo, then Brazil's foreign minister, and Raphael Azeredo, a senior climate negotiator. US-Brazil relations had been frozen for months after news reports in September 2013 that the US National Security Agency had tapped the phones of President Dilma Rousseff and top aides.[3] When my office called in the spring to set up meetings for me in Brazil with Luiz and Antonio, they got nowhere, and neither did I when I sent Luiz an email. This was a problem because the Brazilians were always important, and I wanted to see them. And then one day I realized that the answer was soccer. I sent a new email to Luiz saying that I was coming to Brazil for the World Cup, so perhaps we could meet socially while I was there. He answered in minutes and then arranged the dinner at Antonio's house. Since this was an unofficial visit, I brought my fourteen-year-old unofficial special assistant and first-class soccer player, Zachary Stern (on my own nickel of course), and he was a big hit at dinner after the extended and productive conversation on climate negotiations shifted to an extended conversation about soccer. And as a man of my word, Zachary and I attended two World Cup games.

* * *

As of the end of the day Wednesday, December 10, the draft text for a Lima decision was in a precarious state, having shape-shifted during the week. Artur Runge-Metzer (European Union) and Kishan Kumarsingh (Trinidad and Tobago), who had continued after the Warsaw COP in their roles as cochairs of the negotiations, had issued a slim new draft decision text on Monday morning, December 8. When they convened the parties that Monday, they agreed to walk through the text paragraph by paragraph and allow the parties to propose whatever alternative language they wanted, projected on a screen for all to see. By Wednesday evening, the working draft had ballooned to sixty pages—which was exactly what Artur and Kishan had expected. They knew that they would be criticized if they didn't respect the "party-driven process," so they gave negotiators free rein until an inevitable and unworkable mess had been created, at which point, with the clock running down, the parties would ask them to assert themselves and produce a concise draft. And that's what happened.

But there was a glitch. On Thursday morning, their new draft had mysteriously appeared on the UNFCCC's website before it was ready for release, lingered there for fifteen minutes, and then abruptly vanished. But fifteen minutes was long enough for people to see it, and for rumors and suspicions to start racing around the conference hall. Whatever smooth choreography Artur and Kishan had planned for introducing their new draft was in tatters, and the all-important UNFCCC commodity of trust had taken a hit. The rest of Thursday was mostly consumed in various kinds of informal consultations, as the formal "contact group" session of all parties was suspended.[4]

Late Thursday night, at the direction of COP president Manuel Pulgar-Vidal, Peru's environment minister, the cochairs issued a new draft text, largely similar to the phantom text from that morning. On Friday, Pulgar-Vidal conducted a long series of consultations, supported by two ministers he had appointed as "facilitators," Vivian Balakrishnan of Singapore and Tine Sundhoft of Norway. Sue and I met at different times both with the Peruvians and the facilitators, underscoring our views on differentiation, the scope of INDCs, and establishing a fair process for *ex ante* review. Consultations continued well into the night, with delegations lined up outside Pulgar-Vidal's temporary office. Shortly after 2:00 a.m., Pulgar-Vidal and the cochairs brought the full contact group back into session and circulated another new draft text, dated December 12, plainly designed to be a workable compromise.

Artur, in the chair, proposed that the parties take a half hour to review the four-page text, after which the session would resume as a formal closing plenary to adopt the decision. This was a mistake, and the parties predictably cried foul about being given so little time. At around 3:30 a.m., with no viable possibility of finishing that night, Artur and Kishan adjourned the session to give countries time to consider the text, meet in their various groups, and reassemble on Saturday.

* * *

Saturday morning the proceedings resumed at around 10:30 a.m. as a formal closing plenary. But the ill will from the night before was still on full display. Malaysia's chief negotiator, Gurdial Singh Nijar, a passionate lawyer and orator speaking for the LMDC group, assaulted the text, starting off hot and getting hotter: "We come from different starting points. . . . Many of you colonized us. . . . This is why there must be a differentiated treatment between Annex 1 and non-Annex 1 countries."[5] The current text, he said, failed to do that. He also denounced the lack of balance between mitigation and the other issues as well as the remaining language about an *ex ante* review. And he denounced the ways he felt the cochairs had ignored LMDC efforts to compromise. At the end, addressing Artur, Nijar said, "There is a world out there that is different from your world. There is a poor world, a disenfranchised world." Others joined in. Tuvalu's combative negotiator, Ian Fry, a climate veteran, attacked the text's failure to include loss and damage, an issue of singular importance to the island states. Saudi Arabia argued that the text fell short in terms of both differentiation and adaptation. Sudan, on behalf of the Africa group, also protested.

Many parties supported the text, including the European Union, Switzerland, Mexico, Chile, Belize, and the Marshall Islands, with powerful remarks by Minister de Brum: "I have fought for my country at the United Nations for three decades, beginning with our independence. I know the level of discomfort that is sometimes required to keep things moving forward. So even if we are unhappy about parts of the text, I am prepared to put aside those differences, to keep working on our new agreement, because without moving forward now, and without success in Paris, my country is on the line." Extended applause followed.

But during the next forty minutes, many others, including India and China, piled on against the text. China argued that it was unbalanced and did not reflect the principles of CBDR. Shortly after 12:30 p.m., worried that these strident broadsides against the text might be leading to a stalemate, I took the floor, intending my remarks to be a wake-up call to the full conference. I implored delegates not to lose sight of what was on the line: "The success of this COP here in Lima is at stake. The success of next year's COP in Paris is at stake. And I think the future of the UNFCCC as the body to address climate change effectively is at stake." I urged negotiators to be mindful of what we could achieve if we succeeded in Paris—that for the first time we would establish a stable, durable agreement with legal force more ambitious than ever before, even if not yet enough; applicable to all in a genuine, not formalistic manner; differentiated so that the interests of all developing countries are protected; built on a foundation of rules-based accountability; emphasizing adaptation; calling for large-scale financial assistance; and sending a potent signal to the global community—from boardrooms to civil society—that the governments of the world mean business on climate change. I ended by saying, "I urge you to look at what's at stake, don't throw away what we can achieve, and let's make progress now." It's always hard to tell, but my remarks appeared to have had an impact, based on the reactions I heard from people in the hall.

Pulgar-Vidal suspended the meeting at around 1:30 p.m. to undertake further consultations. He and his team met with parties through the afternoon and evening on Saturday. They circulated the final text shortly after 11:30 p.m. as the closing plenary convened. This text deleted the already weak *ex ante* review paragraph, leaving only a directive to the secretariat to prepare a synthesis report on aggregate INDCs by November 1, 2015. The draft also weakened the paragraph on providing clarifying information, changing the requirement for all parties to provide information ("shall") to a request ("may, as appropriate"). On common but differentiated responsibilities, the text included the US-China formulation as Xie and I had agreed. On finance, the text also further weakened language about expanding the donor base, simply recognizing "complementary support by other parties." And the final text added paragraphs in the preamble about adaptation as well as loss and damage, responding to a demand that various opponents of the December 12 draft expressed on Saturday morning. China clearly had a hand in a number of these changes given its leverage with Peru as the only party capable of delivering the opponents of the December 12 draft. The COP adopted the final text, the "Lima Call for Action," at around 1:30 a.m. on Sunday morning, December 14.

Postmortem

The Lima COP, though unnecessarily bumpy, largely succeeded in doing what it had to do in preparing for the final year of this four-year negotiation. In her short wrap-up piece for *E&E News*, Lisa Friedman quoted China's Su Wei, saying, Lima "lays solid ground for meetings next year leading to a successful meeting in Paris"; India's Prakash Javadekar, noting, "We got what we wanted"; the Africa Group's Seyni Nafo, remarking, "It's not too bad. . . . It's something we can live with"; and me, observing that it was "a good outcome . . . that will get us started on the way to Paris."[6] We all had different perspectives, but knew the crucial hurdle had been cleared and that the sprint to Paris could begin.

There were several important takeaways from Lima. First, COP 20 completed the important work begun in Warsaw to bring alive the core phrase of the Durban Platform; "applicable to all parties." Durban made the conceptual breakthrough to endorse that principle. But first Warsaw and then more definitively Lima moved to make that principle operational by calling on all parties, in an undifferentiated manner, to submit INDCs to the UNFCCC secretariat with stipulated content (mitigation pledges), clarified by stipulated information, within a stipulated time frame designed to allow a consultative period (though not process) in which countries, civil society, analysts, and the press could critique INDCs.

Second, on the endlessly disputed nature of differentiation between the two categories from 1992 (Annex 1 and non-Annex 1), the side against a firewall made important progress. Developed countries, staying united, stared down the demand for a bifurcated information annex that would have badly undermined our negotiating position for Paris. And the US-China formulation on CBDR was adopted by the COP, thereby effectively becoming part of the core CBDR/RC phrase itself. The breakthrough on language had come in the negotiations over the US-China Joint Announcement, but importing those new words into the Lima decision, subject as it was to the COP's quasi-consensus rule, was another critical step forward. Laurent Fabius, the 2015 COP president, described the new wording as "dramatic, new and very useful."[7] Christiana Figueres called it an "important breakthrough."[8] The always careful and circumspect *Earth Negotiations Bulletin* said in its summary of COP 20 that the "outcome document, arguably, shifts the wall of differentiation.'"[9] Michael Jacobs from the United Kingdom thought the combination of a new CBDR formula with no bifurcation effectively killed the firewall because it signified that CBDR now meant something *different* from the firewall—a new kind of differentiation.[10] The *Guardian* editorial the day the Lima COP ended, shrewdly describing the conference as "a skirmish before the real battle," said, "There are no longer two categories of nations, the developed countries, which must cut emissions, and the developing countries, which need not. What the negotiators called the firewall between the two has been breached."[11] Elliot Diringer of the Center for Climate and Energy Solutions commented, "The handwriting is on the famous firewall—it's coming down." I wasn't ready to go quite as far as Michael, the *Guardian*, or Elliot because I knew the firewall proponents would be back with a vengeance in 2015. But it was clear that we had taken a significant step forward.

Third, Lima demonstrated the power of the US-China Joint Announcement with regard to both its impact on countries' belief that Paris would succeed and China's own commitment to getting a deal done. The United States and China were not the G2. We still disagreed on pivotal points, and many other countries played crucial roles. Moreover, Minister Xie would still be thrilled to land an agreement with as much bifurcation as possible. But we were now in this together; Xie knew that and embraced it. The line from the Joint Announcement—"Our presidents resolved to work together closely to address major impediments to reaching a successful global agreement in Paris"—meant what it said.

9 Paris

> The best chance we have to save the one planet we've got.
>
> —Barack Obama

From 1992, when the UNFCCC was adopted in Rio de Janeiro, to 2015, every effort to put in place an effective, operational climate regime failed. Now the squabbling nations of the earth had one more chance to get it right. And at last the stars seemed to be aligning. Countries had already accepted important elements of an agreement, such as the nationally determined structure for emission targets. The US-China Joint Announcement gave China a greater need to get a deal done than ever before and instilled in countries around the world a new confidence that a deal would get done. Countries knew that the clock was running down on the tenure of a much admired and deeply committed US president. The UNFCCC secretariat was commanded by the unstoppable Christiana Figueres. And the French COP presidency, led by Foreign Minister Laurent Fabius and his top aide Laurence Tubiana, had the skill, finesse, and expertise to steer a new agreement to port. In short, if not now, when?

Yet it would be a delicate dance to complete not just any deal but also the right deal, a strong enough deal, knowing that it only ever takes a handful of dissenters to block or water down a climate agreement. And countries were nervous about how their own priority issues would be treated in the final outcome. The most pervasive anxiety revolved around the differentiated treatment of developed and developing countries. The LMDC were concerned because their highest priority was to defend the firewall, but a great many other developing countries were uneasy about the nature of a whole new climate architecture beyond the bifurcated world they were used to. And countries were anxious about a range of other issues as well—about whether the ambition of the new agreement would be equal to the task of containing climate change, about the legal form of the new agreement, and about adaptation and financial assistance and loss and damage. So, countries approached this gun lap negotiating year with both a greater than usual measure of confidence and a sense of well-founded apprehension.

That France held the presidency for COP 21 was opportune. The French had a long tradition of skilled diplomacy. Fabius, tapped to be the COP president, was an esteemed leader who had been in and around government for decades, and at one point, was the youngest prime minister in the history of France's Fifth Republic. He was the living image of the smooth, logical French Cartesian—an "Énarque" as the French call the graduates of the elite École Nationale d'Administration, France's training ground for future Olympians. He had stature, intellect, and a sense of command. He was a statesman who could sit comfortably with leaders all over the world. And he made climate change his central priority for 2015, underscoring the importance of a Paris Agreement in all of his meetings.[1] Laurence, an old friend of mine, was his chief adviser. Diminutive, mild mannered, and soft-spoken, with a shock of white hair, she was an accomplished academic, activist, diplomat, and climate veteran. She understood the mechanics and culture of climate negotiations, knew what made different countries and groups tick, was the right partner for Fabius, and assembled an excellent team of experts.

As for the United States, we began the year with more credibility and leverage than ever as a product above all of President Obama's intensive climate activism at home and abroad in his second term—an activism that other countries saw and appreciated. Meanwhile, Secretary Kerry, a climate warrior for twenty-five years, exhorted countries all over the world about the need to land an important climate agreement, and played a pivotal role in promoting US-China collaboration. Inside the White House, Brian Deese stepped in to replace John Podesta, who left in January 2015 to advise Hillary Clinton in her presidential run. Brian, smart as a whip, strategic, and wise, was in his late thirties and looked even younger, with a wispy beard and a smile never far away. Indian prime minister Narendra Modi's chief economic adviser, Arvind Subramanian, referred to Brian during a June dinner I had with him in New Delhi as "the youngster in the White House who does climate," while Minister Xie, after asking about Brian's age, noted that he was younger than Xie's own son. But nobody doubted him within five minutes of meeting. He complemented his intelligence with an instinct for leadership and an effortlessly collegial manner. With the Paris Agreement center stage, I saw Brian at the White House most every week and couldn't have asked for a better partner.

* * *

Our objectives for the year were clear. We wanted an ambitious, durable agreement that was genuinely applicable to all with a modified form of differentiation but no firewall. We wanted strong accountability measures such as obligations for countries to submit NDCs and ratchet them up every five years, and a single system to review countries' progress toward meeting their NDC targets. Of course, we had to have an agreement with a hybrid legal character, partly legally binding and partly not. And we sought workable provisions on financial assistance, adaptation, and loss and damage. The trick would be figuring out landing zones on tough issues that gave a wide coalition of players enough of what they needed. We pursued these objectives through a year packed with diplomacy of all sizes and shapes, with a wide range of countries from large to small, from allies to those who broadly disagreed with us.

China: The Joint Statement

The first MEF meeting we hosted in Washington in April illustrated the two sides of China in the aftermath of the Joint Announcement. The most important development occurred in a side meeting I had with Minister Xie. We agreed to use the occasion of President Xi's planned visit to the White House in September as a spur to negotiate a new presidential joint statement aimed at nailing down as much agreement as possible on key issues for Paris. This effort played out in five negotiating sessions in May, June, July, and September in Berlin, Beijing, Luxembourg, Paris, and Los Angeles.[2] That was one side of China. The other was visible in the MEF meeting itself. After I said that a new agreement would need to include a common transparency system in which all countries report and are reviewed on the implementation of their targets, Xie said that in Paris, only broad principles should be agreed on and more detailed "transparency arrangements" should be done "afterward." This was unacceptable. We saw transparency as a crucial part of any Paris Agreement, the essential means by which countries can be held to account for doing what they pledged to do. And a strong, legally binding transparency regime would bolster the credibility of the overall agreement, particularly for those troubled by nonbinding emission targets.[3] Plus, transparency was the issue where the rubber would meet the road on bifurcation. The handling of transparency presented a clean binary: there would be one set of common rules or there would be two sets of rules. The new paradigm or the old. Cañete, the European Union's commissioner, shared our commitment to a strong, unified, binding transparency system. "It does not make sense to run two separate MRV systems in parallel," he said. "Those with less capability will be allowed to supply less detail. All parties need to be accountable for their commitments." At our final MEF meeting of the year, he said simply, "We are talking about the most important element of the Paris Agreement."

My take on Xie's posture was that he thought China could better preserve a version of the old firewall if the operational elements of transparency were negotiated later, in a less closely watched setting, away from the klieg lights of Paris. This conflict—between those favoring a minimal agreement and those favoring a robust one—was a theme played out all the way to the last days of the Paris COP. Yet that Xie, on the same day that he pushed for minimalism on transparency, could say to me, in effect, "Let's do a new joint presidential statement so we can pursue common ground on tough issues," captured the ying and yang of the US-China climate relationship that year: coming together in a high-level, cooperative endeavor to secure a Paris Agreement and pulling apart in our different conceptions of what the Paris Agreement should actually say.

* * *

In Berlin in May, right after the end of the 2015 Petersberg Dialogue, Xie and I launched our "joint statement" discussions during a brief meeting. We agreed to explore the idea of a long-term, aggregate emissions goal, the periodic update of country emission targets, transparency, differentiation, and legal form. We had our first extended conversation in late June in Washington on the margins of the annual US-China S&ED meetings. In the course of this session, Xie pushed a hard line on differentiation, and to my considerable surprise, questioned the widely accepted notion that there would be subsequent rounds of NDCs after the initial round of NDCs expired in 2030. After listening to him for a while, I leaned back and said that if he held firm to these positions, there would be no agreement that the United States could join. He briefly reiterated a couple of his points and then shifted gears to say, "The reason I enjoy my discussions with you and have built mutual trust is that I admire and appreciate your candor, and I can carry on in-depth discussions with you. We are free to share our bottom lines. That is the foundation for us to find a solution. We have a great deal of mutual trust. I don't like some other friends who will talk to your face in one tone and do another. I don't like that at all. We have worked together for seven years. Building on that basis, we can find a solution." These were nice sentiments, and I appreciated them. But we still had a long way to go to complete a joint statement for our presidents, not to mention an agreement in Paris.

We met again on July 17 on the margins of a MEF meeting in Luxembourg and again made little progress. I pushed him hard on successive rounds of NDCs since the notion of a Paris Agreement that would, or even might, end in 2030 seemed preposterous. I said there was wide support for a long-lasting Paris Agreement with subsequent rounds of NDCs, and such a structure was crucial to building confidence and trust, strengthening the message to business and civil society, and bolstering the ambition of the agreement. He crossed his arms while listening to me—a sure sign that he was getting agitated—and said that he agreed there should be more rounds of NDCs in general, but that "we should not prejudge"—meaning we should not nail that down now. I had different theories about why he was digging in his heals here. He was unhappy with the recent G7's embrace of a goal to reduce emissions 40–70 percent below 2010 levels by 2050, and I thought he might want to resist accepting successive rounds of NDCs until he knew that kind of long-term goal was off the table.[4] But whatever the reason, I had to move him off that position. I pushed him hard on this the next day in our MEF meeting so he could see that China did not have much company on this issue.

He was also dug in on bifurcation. Xie said that for "implementation reviews"—an aspect of transparency—"we still insist on different systems to review developed versus developing countries." And a few minutes later, he added, "The commitment of a developed country in regard to its NDC should be made at a certain level and developing countries can do as much as they can do." This was such a clear distillation of exactly what the

United States would never accept that I answered, "But that is the difference between an agreement in Paris or not."

It was interesting to me that Xie's unreconstructed invocations of the firewall came after the 2014 US-China joint announcement added important new language that softened the classic differentiation phrase—"common but differentiated responsibilities and respective capabilities *in light of different national circumstances*." But when we next met on September 6 in Paris, it became apparent that he actually wanted to modify that new joint announcement formulation of CBDR. He pushed me to add language to the 2014 formulation so "our joint statement can reflect inclusiveness." He wanted to add words like "on the basis of equity" or say that the new agreement will be meant to "enhance the implementation of the Convention" (another way of insisting on the continuing vitality of the firewall). I explained to him quite candidly why we could not do that. I said that we didn't accept that the 1992 categories should be the basis for a new agreement. I acknowledged that our 2014 formulation was a bit ambiguous—that we read it to suggest a differentiation across a spectrum rather than between one group and another, and noted that it conveys a sense of change over time because national circumstances change over time. Nonetheless, I said that I knew that some countries claimed that the new formulation was just the plain old CBDR (although there would of course have been no reason for a new formulation if there were no intent to change the old formulation). I told him the new sentence represented a delicate balance that the two Su(e)s worked out last year after hours of discussion, and the suggestions he proposed now would distort that balance and so were unacceptable to us.

With that, Xie moved on. But it is worth underscoring that the kind of constructive ambiguity in negotiations that Xie and I were discussing does not mean that the two different ways to read a given phrase have equal weight. More commonly, one reading is regarded as more compelling than the other and over time wins out. And we thought the reaction to the new CBDR formulation in the Joint Announcement by objective, outside observers well-versed in the nuances of climate diplomacy weighed decisively in our favor.

* * *

For the United States, the most important element of the new US-China joint statement was the short paragraph on transparency, which we negotiated in our last session on September 15 in Los Angeles on the margins of the first US-China cities summit, itself a product of the 2014 Joint Announcement. Xie and I went back and forth on the issue, with me refusing to accept a bifurcated transparency system, and Xie refusing to call it "single" or "unified." When we reached as much clarity as we could, we dispatched the two Su(e)s to work out the precise language, mindful that we had to finish the joint statement in Los Angeles since the summit at the White House was starting just ten days later. The words they brought back an hour or so later were that Paris should establish "an enhanced transparency system" that would include reporting and review of both action and financial assistance, and "*should provide flexibility to those developing countries that need it in light of their capacities.*" This language was a breakthrough. "Flexibility" in the way countries needed to carry out the dictates of a new system of reporting and review was the means of providing differentiation, but now China was agreeing that it would not be accorded to all developing countries simply because of their developing country status but only *to those that need it* because they lacked *capacity*. In short, Sue had again worked her magic. That short sentence moved us an important step forward.

Brazil

On May 13, at the Brazilian embassy in Washington, I met with my old friend Luiz Figueiredo, then Brazil's ambassador to the United States, to discuss President Rousseff's upcoming visit to the White House, some twenty months after she canceled her original Washington visit in protest over the phone-tapping affair. Luiz told me that Rousseff wanted climate change to be a prominent focus of her June visit, and hoped for a joint statement of the kind that Presidents Obama and Xi announced six months earlier. Luiz also explained that Izabella Teixeira, the environment minister whom I first met in 2009, was a good friend of Rousseff's and would be taking a more prominent role in the negotiations. This was all good news.

Sue and I worked intensively with Luiz in May and June to craft a joint presidential statement that was strengthened by Rousseff's decision, after a call from Vice President Biden, to spell out Brazil's post-2020 plans during her visit. All of this amounted to strategic progress for us. Brazil was an important player—a member of BASIC, influential, and diplomatically skilled. We were not allies in the negotiations, but a repaired relationship between Presidents Obama and Rousseff, with a focus on climate change, the active participation of Minister Teixeira, and the renewed climate engagement of Luiz all augured well for our ability to work together in Paris.

India

I traveled to New Delhi in June, focused particularly on transparency. As expected, I made little progress with Minister Javadekar and his team. We had arranged meetings with a number of other important and more pragmatic officials in the Indian government, including chief economist Arvind Subramanian and Ashok Sharma, a senior adviser to Prime Minister Modi, to convey the message that the United States was ready to be a productive partner on climate and clean energy if we could find a way forward in the negotiations. Brian Deese took his own trip to India in September, conveyed a similar message to Foreign Secretary Subrahmanyam Jaishankar, and delivered a letter from President Obama to Modi. Their relationship was the most important card in our hand. Modi's September 2014 visit to Washington had gone well. He appeared to have been particularly moved when President Obama took him on a visit to the Martin Luther King Jr. memorial. He told the press, "He took out a lot of time. . . . We were together yesterday and today for quite some time, and today in fact he took me around, and with such ease and such humility." He thanked the president "from the core of my heart."[5] Just two months later, he followed up with an invitation that Obama promptly accepted to celebrate India's Republic Day parade in New Delhi on January 26, 2015. That same year, he announced an enormously ambitious target to increase India's installed renewable energy capacity from 36 gigawatts to 175 by 2022.[6] So there was reason to hope that the prime minister and his pragmatic advisers would, in the end, exert a constructive influence on India's ultimate posture in the negotiations.

Small Islands

The small islands, whose vulnerable status gave them a special moral authority, cared little about defending the firewall for the benefit of major developing countries, but cared passionately about two issues: having loss and damage included in a new Paris Agreement, and establishing a goal of holding the rise in the average global temperature to 1.5°C above preindustrial levels rather than "below 2°C."

On loss and damage, we recognized their concern about the exposure they faced even if they did everything possible to make their economies resilient, but the issue was politically charged. If loss and damage language in a Paris Agreement could be read to imply that developed countries were liable and owed compensation for potentially enormous worldwide damages, it would hand climate opponents in the United States a potent weapon to oppose a new agreement.[7] In short, we needed to find a solution workable for both of us. And I wanted to find that solution before the Paris COP. Loss and damage was an emotional issue that had nearly derailed the Doha COP in 2012 and had loomed as a potential stumbling block in the final plenary at the Warsaw COP in 2013. I met frequently during the year with Ministers Tony de Brum of the Marshall Islands and Jimmy Fletcher from Saint Lucia, often with Sue, Trigg, and Christina Chan. In early September, I dispatched Trigg to Papua New Guinea to attend a gathering of leaders of the Pacific Islands Forum, where he met at senior levels with most of the eighteen island members. I did one-on-one meetings in New York with President Anote Tong of Kiribati and Prime Minister Enele Sopoaga of Tuvalu during the September UN General Assembly meetings. And I joined our UN ambassador, Samantha Power, also during the UNGA meetings, at a roundtable of the Pacific Island Forum. This diplomacy built relationships and developed a common understanding of what each side needed in a solution so that by the time we had our last meeting before the Paris COP in early November, we had a good idea of what the landing zone was likely to be. And our diplomacy also helped align us on key elements of ambition that proved pivotal in Paris.

Luxembourg and Paris

In July, we visited Luxembourg and Paris for two major meetings and a series of bilateral conversations. At a rough midpoint of the year, these sessions provided a useful snapshot of where things stood on pivotal negotiation issues and the overall state of mind of negotiators as the Paris COP drew closer. In a broad sense, I came away from the trip somewhat encouraged. Most countries—twenty-six attended our MEF meeting in Luxembourg, and fifty to sixty came to the meeting hosted by the French in

Paris—appeared to assume that a deal would get done. And that included some, like Bolivia, that were members of the hard-line LMDC and ALBA groups. When I met with Bolivia's lead negotiator, Rene Orellana, during the French meeting, he told me that the US-China Joint Announcement had given his country confidence. He said that its friend China had explained the Joint Announcement to Bolivia, and that the announcement was the key to bringing the ALBA and LMDC groups along. Rene also said the pope had made a difference with his climate change encyclical, *Laudato Si*, telling me that his president had decided to increase Bolivia's climate action "because the encyclical demands it." The tone of our conversation was illuminating because of how optimistic and cordial it was. The sense of the United States as the bad guy, quite common in the rhetoric of ALBA countries, was nowhere to be found. In addition to the impact of the Joint Announcement, I expect that President Obama's activism at home and our own conduct in the negotiations played a part as well, as did a trip that Trigg took there in April.

I also was struck by what seemed a new day with Brazil, owing, first, to President Rousseff's successful visit to the White House, including the joint presidential statement on climate and more active role being played by Minister Teixeira, on display in the French ministerial meeting, where she acted as a cofacilitator. This was a welcome development since she was a climate activist who wanted to make a difference and had Rousseff's ear. And Luiz's likely participation in Paris meant that along with Izabella, we would have multiple channels of communication beyond their hard-hitting climate lead, Antonio Carvalho.

* * *

Following a suggestion EU commissioner Cañete and Minister Xie made to me at our MEF meeting in April, we engaged outside facilitators to lead discussions at our July and September MEF meetings in order to better explore common ground on key issues. This worked particularly well in the transparency session where Kwok Fook Seng, Singapore's lead negotiator, was not only exceptionally talented but had also understood the thinking of the negotiators of the MEF countries, since he facilitated this issue during the various Bonn intersessional meetings all year. Fook Seng would make an interjection here, a small summation of a point there, taking the sharp edges off a comment and steering in such a way as to reveal possible areas of compromise, all with a completely nonjudgmental tone that led no one to think he was against them. Clearly mindful of where he thought the ultimate compromise lay, he gently led the process so others might start to see a way forward. And the truth was that except for India, China, and Saudi Arabia, which were still holding out to continue a bifurcated system, there seemed to be gathering support in these meetings for a common system with a good deal of flexibility to account for countries' different capacities as well as some kind of transition period at the beginning.

* * *

The French managed the Paris consultation well, focused on the key issues of differentiation and ambitious action to reduce emissions. But in the final plenary gathering, focused on ambition, I was concerned by a worry pervading the room that Paris would be deemed a letdown because country targets would fall short of putting us on track to meet the iconic below 2°C temperature goal. Minister Fabius was also concerned and talked of the need to avoid a "misunderstanding in the public of what this agreement means." I took the floor after Fabius, intending to rally the troops, arguing that if we otherwise produced a good agreement, we should be firm and confident about Paris ambition. I made five points: we will have delivered a large number of emission targets (NDCs), not perfect, but a good sign of commitment, and by themselves, a big step forward on ambition; there will be a built-in ratchet mechanism for countries to strengthen their targets regularly—we hoped every five years; there will be a strong long-term goal guiding the whole global effort; countries will be called on to prepare long-term strategies for decarbonization by 2050—an idea Laurence first proposed; and all of this will be buttressed by the intensifying engagement of "nonstate actors" such as cities, states, provinces, businesses, and civil society. I said that if you put this package together, we'll have a strong message to convey in Paris, with no reason to hang our heads. This pep talk actually made a difference.

* * *

I had a short pull-aside conversation with our Saudi friend, Khalid Abulief, during the July consultation in Paris. The Saudis are always important, and they were a big focus for us in 2015. Khalid was a smooth, able, and gracious Saudi diplomat, and we had made a point of spending time with him after he replaced his unreconstructed predecessor, Mohammed al-Sabban. Kareem Saleh, my Arabic-speaking chief of staff, and I visited Khalid in Saudi Arabia in 2012, and Sue and I talked with him often, including over dinner in May 2015 at his favorite Chinese restaurant in Paris. The Saudi position on key issues, including transparency and differentiation, was quite problematic for us, but in this short tête-à-tête, I didn't want to discuss the issues. I just wanted to convey a clear message: that a strong climate agreement in Paris was a high priority for us, all the way up to President Obama, so it was going to be critical for the United States and Saudi Arabia to work closely together to get a deal done, and important that Saudi Arabia not get positioned as standing in the way of a deal.

* * *

Sue and I also had a second meeting with Egyptian environment minister Khaled Fahmy and his chief negotiator, Mohamed Khalil, after a long session we had with him at the Egyptian embassy in Paris in early May. Egypt played an unusually influential role in 2015. It hosted the annual African Ministerial Conference on the Environment, (AMCEN) and acted in 2015 as coordinator of the Committee of African Heads of State and Government on Climate Change. Trigg and Karen Johnson (a colleague of Sue's from the Office of the Legal Adviser at State) attended a March session of AMCEN, meeting with negotiators from a dozen countries. Plus Egypt was a member of the LMDC group. For all of these reasons, Sue and I met regularly that year in Paris with Minister Fahmy and Khalil. Their central concern was the move away from the firewall, complaining that developing countries were being offered nothing for giving that up. We listened and patiently pushed back on this perception, and though we never changed their minds, I thought the hours we spent together were more than worth it. They heard our arguments. They could see that we cared about their views. And we liked each other. All of these factors might well prove valuable at the Paris COP as the final compromises were sorted out.

* * *

Sue and I met in Paris with Angola's young, lead negotiator, Giza Gaspar-Martins, who served as the head of the Least Developed Countries group. I liked him right away. He was one of those people who naturally integrate policy and politics, and I thought that if he had been an American,

I could have seen him working on Capitol Hill or in the White House. Significantly, he was not stuck in the old UNFCCC orthodoxies, and was not afraid to stand up and speak out. He strongly supported a global temperature goal that included 1.5°C, not just 2°C, and financial support for poor countries, but told us that LDCs were ready to move beyond "binary differentiation"—meaning the firewall. And he had made the same point out loud during the recent MEF meeting, arguing that there should be a single set of transparency rules, not two, and that this issue needed to be resolved at the Paris COP so that we did not find ourselves in a debate over fundamentals in the years after the COP.

* * *

In a bit of serendipity, I also started to make friends with Cuba, another ALBA country. As it happened, the United States and Cuba had reopened their embassies in Washington and Havana, respectively, on July 20, the first day of the French consultation. The seating arrangements—alphabetical in French—put the United States (les États-Unis) quite close to Cuba, so I introduced myself to the Cuban delegate, Deputy Minister José Fidel Santana Nuñez, said that this was a big day for our two countries, and suggested that we should talk during a break in the action. We did that during a coffee break and had a cordial exchange. I said that I'd like us to talk in more depth and perhaps I could visit Cuba, which Nuñez encouraged. Sue and I ended up doing that on a fascinating and productive day in late October, when we met with a number of Cuban ministers and had enough time to see a little of Havana. Our short trip was valuable most of all because, as so often happens, it changes the game a little when you reach out, visit someone on their home turf, and listen to what they have to say. Policy differences don't disappear, but the nature of a negotiator's opposition or disagreement can soften. And that can matter.

* * *

In the run-up to our last MEF meeting of the year, late September in New York, we talked internally about how to sweeten the pot on transparency. Dan Reifsnyder told us about a system used in the Montreal Protocol to provide in-country transparency expertise to help developing countries with their reporting duties. Dan said that this kind of program could be established and operated with a modest amount of money, which developed countries could easily provide. This was another of Dan's practical ideas, and I went for it right away. We gave it a name—the capacity-building initiative for transparency, and I introduced the idea in the opening session of our MEF meeting, where it was well received. It was eventually agreed to in Paris and is now run through the Global Environment Facility.

An October Surprise

At the year's third intersessional meeting in early September, where negotiators were still working off an unwieldy, eighty-five-page compilation text, they agreed that the cochairs of the negotiation—our own Dan Reifsnyder and Ahmed Djoglaf from Algeria—would produce a short text in the first week of October to be discussed at the next intersessional meeting, October 19–23 in Bonn, after which it would form the basis for the negotiations at the Paris COP. That was the plan. But as Mike Tyson famously said, everyone has a plan until they get punched in the mouth. The punch came in Bonn.

The cochairs issued their draft text on October 5. It was the first, short, well-ordered draft that looked as though you could actually negotiate from it in Paris. Up until then, many, including the French, had been wringing their hands because the negotiating process seemed incapable of making progress. Now, at least, we had a concise, nine-page document.[8] But in short order the draft was lambasted as favoring developed countries. By the time the meeting formally began on Monday, October 19, South Africa's Nozipho Diseko, chair of the G77, joined by Sudan, chair of the Africa Group of Negotiators, refused to proceed unless the new text was opened for "surgical" insertions, which turned out to be not very surgical. By the time the Bonn meeting ended, the spare, nine-page text had swelled to thirty-one pages, littered with options and brackets.

And Diseko played hardball. She claimed that the climate negotiations were an example of the subjugation of developing countries by developed. She claimed it was "like apartheid," with poor countries getting ignored. And she demanded unity among the G77 subgroups, despite the significant differences among them, thereby muting progressive voices and amplifying the voice of the LMDC. A French newspaper, *L'Echo*, wrote an article about her headlined "La Diva du Climat."[9] Thus although my team had reported back to me that there were signs of convergence on many issues in side conversations, the larger dynamic had turned bitter. I saw the potential for a noxious narrative to develop, attacking developed countries for trying to abandon differentiation and shirk their financial obligations.

A November Response

I thought we had to cut that narrative off by making it clear that financial assistance would flow if we collectively produced a good deal, that developing countries had nothing to fear from the flexible but nonbifurcated system we sought, and that we could not build an ambitious, long-lasting agreement on the foundation of a firewall with rapidly rising developing country emissions already accounting for over 60 percent of the global total. To produce the kind of ambitious agreement that the United States wanted, we needed progressive voices to be loud and insistent the way they were during the 2011 negotiations for the Durban mandate—or even louder. Pete Betts said it most succinctly at a US-UK bilateral in New York in September: "If it's us against the developing countries, we lose; if it's us plus vulnerables and progressives, we win."

I looked at the upcoming pre-COP meeting in Paris on November 9–10 as an opportunity to begin recovering from the ill wind that started to blow in Bonn. I had three steps in mind. First, I reached out to influential developed and developing country friends to propose that we gather for a small dinner in Paris a day or two before the pre-COP to discuss reviving the progressive coalition from 2011—but this time with the United States included. Second, I wanted to meet with Diseko to see whether we could calm things down and reach a better understanding, but though I offered to meet her in Johannesburg, have dinner in Paris, or even just do a short pull-aside conversation at the pre-COP meeting itself, she appeared to have no interest. Third, I wanted to send a signal at the pre-COP that the United States would support mobilizing $100 billion per year beyond 2020 if the right conditions were met.

This last bit was not what I had argued at a donor finance meeting in the summer. Then, when a number of my European colleagues said that continuing the $100 billion per year after 2020 was effectively already in the pocket of developing countries, I took exception, saying we were in a negotiation and continuing the $100 billion was a top priority for many countries that might oppose us on other issues, so we had to treat it as a card in our hand. But now, after Bonn, we were facing different circumstances and needed to rebuild a progressive coalition. Timing matters, and it is in the nature of negotiations that tactics sometimes shift. That said, not everyone in our own administration agreed. At a White House principals meeting in the Roosevelt Room a few days before I left for Paris, my colleague Paul Bodnar, formerly the lead finance negotiator on my team, now at the White House, opposed showing any movement in the US position until it was clear the donor pool would expand beyond developed countries alone and that any extension of the $100 billion commitment would be time limited. A number of people supported Paul's position. Indeed, I agreed that Paul's conditions were what we would ultimately want in the agreement. But I thought that sending a signal of flexibility now, at the pre-COP, was essential for reanimating the progressive coalition. And I followed my own instincts.

* * *

Sue, Trigg, and I hosted the progressives dinner in Paris on Saturday night, November 7, at La Ferme Saint-Simon two days before the pre-COP meeting itself began. The attendees included EU commissioner Miguel-Arias Cañete, Artur Runge-Metzger (European Union), Pete Betts (United Kingdom), Karsten Sach (Germany), Tony de Brum (Marshall Islands), Andrea Guerrero (Colombia), and Jo Tyndall (New Zealand). I said that in light of the revolt against the cochairs' text in Bonn, the anger expressed against developed countries, and Diseko's insistence on G77 discipline, we needed to reinvigorate the progressive alliance of island states, LDCs, progressive Latins, Europeans, and now the United States. I learned later that a new progressive coalition had already started forming months earlier on the sidelines of the French ministerial consultation in Paris in July among Cañete, de Brum, and Pablo Vieira Samper (Colombia). The group met a few more times on the margins of other events, growing larger from meeting to meeting. So my idea was already in motion, but in convening the dinner, I had added urgency to the mission and began pushing the United States into the coalition. As the dinner was breaking up, Tony, who immediately recognized the strategic importance of bringing the United States into the coalition, invited me to a meeting with progressive countries the next morning—the first time the United States had been invited to such a gathering.[10] At Tony's suggestion, Sue and I attended for an hour, and when it was my turn to speak, I situated the United States squarely on the side of progressives.

* * *

On Friday, a few days before the pre-COP began, Sue and I met with Khalid Abulief, the lead Saudi negotiator, with our focus on transparency. Khalid told us as we got down to business in Paris that Prince Mohammed bin Salman and President Obama had spoken recently and given us "a mandate" to work together. And in fact, we struck an important bargain at that Friday session. Khalid had staked out a position on transparency during the year that opposed the single, flexible set of rules we advocated. In the course of our conversation, Khalid agreed that he could support a new unified system if we could assure him both that "in-country" reviews by expert review teams would be optional and that the Saudis would not lose the right to report on the impacts of so-called response measures—meaning the impact on countries like his of measures taken to transition away from fossil fuels.[11] We agreed to both of those conditions. We would have preferred in-country reviews to be at the discretion of the expert review teams, but did not believe we could have won that fight in any event. And Khalid's desire to continue reporting on the impact of response measures was fine. This new understanding with Khalid was a major step forward given Saudi Arabia's prominence as an opponent of a unified transparency system.

* * *

The pre-COP meeting itself, attended by around sixty countries, focused on ambition, financial assistance, differentiation, and adaptation. In the finance breakout session, I said the United States would favor developed countries continuing to mobilize $100 billion per year in support after 2020 in the context of meaningful mitigation action and transparency by developing countries—the same conditions agreed to in 2009 and 2010. Then I added that since we were in the midst of negotiating a new agreement, the new agreement itself would need to include meaningful mitigation and transparency provisions. I said, by way of example, that if the transparency article of a Paris Agreement called for two sets of rules, one for developed countries and one for developing, we would not regard that as meaningful transparency. Developing countries seemed pleased to hear my conditional willingness to continue the $100 billion.

On the second day of the pre-COP meeting, Sue, Trigg, and I met with de Brum, Fletcher, and their teams to discuss loss and damage. I repeated our position that if the issue were going to be highlighted in a Paris Agreement, it would not be enough to simply avoid talking about liability and compensation. We would need language explicitly stating that the inclusion of a loss and damage article in the new agreement would not imply support of a legal claim for liability and compensation. But I also explained that we were not attempting to prevent developing countries from arguing in court that such a right existed; we just did not want the Paris Agreement to be interpreted as bolstering such an assertion. De Brum and Fletcher were pleased to hear this clarification, and pleased by Sue's overall presentation of what we thought a loss and damage article in the agreement should look like. They said they would discuss our proposal with others in their group. "I think we have gone very far with this," Fletcher said. "I think the landing gear is coming down on the plane, but we do not yet have permission to land."

Sue and I also met with Angola's Giza Gaspar-Martins, the LDCs lead. When I asked him if he thought we'd get a deal in Paris, he said he believed we would, but that we were at risk of a minimalist deal. Giza had his finger pretty well on the pulse of what went on behind closed doors in the G77, so his warning was a concern. And it was entirely consistent with what we had heard from China since the beginning of the year as well as from Indian minister Javadekar, who regularly sought to remind others that not everything (meaning transparency) needed to get done in Paris.

* * *

I met with Minister Xie toward the end of the pre-COP meeting and told him I was a bit concerned that he had not made himself as available during the two-day meeting as I would have expected, adding that I did not appreciate his support for India's defense of the firewall during the last plenary session. Once again, he seemed more amused than bothered. "Are you losing confidence in me?" he asked with a smile. He said, in a way that came across as more endearing than anything else, "I am going to call you younger brother. . . . I think I can say that we should allow all parties to speak freely about what they want. And the COP will not go as smoothly as the pre-COP, but when things don't go well, we should intervene." I smiled and told him he was free to call me younger brother. We agreed to keep in close touch in Paris and not wait as long as we did in Lima to deal with difficulties because in Paris, a much bigger and more complicated event, that might be too late. Sue suggested that we needed "an early warning system," and Xie liked the phrase.

I talked to Xie one more time before the COP, on November 16. I raised two matters: that there had to be successive rounds of NDCs, an issue pretty well in hand, but not locked down; and transparency. Despite the good transparency language in the September joint statement, China had continued to push for two separate sets of transparency rules. Moreover, the South Africans told us that the Chinese had reassured them that they need not worry about common transparency rules—which was not helpful. Knowing the Chinese aversion to having substantive matters elevated to their leaders, I told Xie that President Obama might well raise the transparency issue with President Xi at their bilat on the first day of the conference. Predictably, Xie was not pleased, but countered by saying that if Obama raised the issue, the two presidents would just instruct their negotiating teams to resolve it.

Obama

For President Obama, climate change had by now become personal. He focused on the Paris conference all year long, and climate figured prominently in virtually all of his leader-level interactions, including with Indian prime minister Modi and Presidents Xi and Rousseff. Obama also raised climate change in a speech in Addis Ababa during his four-day trip in late July to Kenya and Ethiopia.

In late summer 2015, Obama, with *Rolling Stone* writer Jeff Goodell in tow, took a trip to a small, endangered Alaskan village called Kotzebue, thirty miles above the Arctic Circle, to dramatize the dangerous impacts of climate change and urgent need to take action. Goodell's interview and commentary was the cover story for the magazine on October 8 and ran under the title "Obama's Climate Crusade." At a speech he gave in Anchorage, Obama said, "If we stop trying to build a clean-energy economy and reduce carbon pollution, if we do nothing to keep the glaciers from melting faster, and oceans from rising faster, and forests from burning faster, and storms from growing stronger, we will condemn our children to a planet beyond their capacity to repair. . . . On this issue, of all issues, there is such a thing as being too late. That moment is almost upon us. That's why we're here today. That's what we have to convey to our people—tomorrow, and the next day, and the day after that. And that's what we have to do when we meet in Paris later this year."

Obama also made calls in the weeks leading up to the Paris conference to Modi, Xi, and South African president Zuma telling them that the lesson he took from his time in Copenhagen—a time he said scarred everyone—was that major players should not arrive at a COP so far apart in their positions. He thought our teams needed to think about ways to bridge our differences. Of course, working to bridge gaps was the logic that guided me and my team all year long, but it was still useful for the president to carry the message directly to his counterparts. Sadly, he found himself talking to French president François Hollande twice on November 13—the first time to discuss the upcoming climate conference, and the second time to offer condolences for the horrific coordinated terrorist attacks in Paris that night that killed 130 people.

The Paris COP

COP 21, held on the outskirts of Paris at the repurposed Le Bourget Airport, opened on November 30 for the two most consequential weeks of climate negotiations since the Rio Earth Summit in 1992. It opened in a spirit rich with hope that success was finally at hand, but tinged with anxiety that the outcome might not go far enough, in the eyes of those committed to high ambition, or might go too far in the eyes of those devoted to the firewall. More than 150 heads of state and government were on hand for the opening—the largest group of leaders ever to attend a UN event in a single day.[12]

My own mood was a mix of cautious confidence and uncertainty. During the year, the United States and our negotiating partners sketched out potential landing zones on tough issues from transparency to loss and damage. The United States and China had reaffirmed our unusual partnership. The United States and Brazil had established a new working relationship on climate change. President Obama and Prime Minister Modi had bonded. Astonishingly, more than 180 countries had submitted their intended NDCs—a sure sign of belief that a deal would be struck. And the French, after hosting three productive ministerial consultations, well-conceived and well run, were competent and ready. So there was a lot going for us. But the unavoidable truth about climate negotiations is that it takes very little to reduce an agreement to a low common denominator. We knew there were major players who were angling for a minimal outcome. Plus it was a matter of record that the last full gathering of the parties in Bonn in October had gone off the rails.

The most contentious issue on the table was still the nature of differentiation. Our side argued for a more supple, noncategorical form as exemplified by both the self-differentiated structure of NDCs and the transparency language in our recent US-China joint statement, which made differentiation depend on a lack of capacity. But the LMDC group was still pressing for as much bifurcation as possible—in the broad, framing articles of the agreement, different transparency rules for developed and developing countries, and various paragraphs on mitigation. The other highly contested issues were the global temperature and emission goals; how the five-year cycles would work, both for countries to ratchet up their individual targets and to take stock of how the world was doing on climate change in the aggregate; what would happen to the developed countries' $100 billion pledge post-2020; and the legal form of the Paris Agreement.

* * *

On Monday morning, I arrived early at Le Bourget and found my way to President Obama's holding room near where his meetings with the Chinese and Indians were scheduled to take place. He was going to be speaking in the Seine plenary hall within the hour, and when I walked in, he was sitting with a few of his White House aides, looking relaxed:

Obama: Hi Todd, how's it going?

Me: It's going to be tough. It's always tough.

Obama: It's always tough. Saving the planet. It's not every day you get to save the planet. And you don't even have a cape.

Me: I left it at home.

In a little while, the president went to the Hall to deliver his remarks. He talked of the climate damage he had seen firsthand on his late summer trip to the Alaskan Arctic, acknowledged the US role in creating the climate problem and our responsibility to address it, and recognized the need to provide resources to help poor and vulnerable countries build resilience as well as cope with the impacts of climate change that cannot be avoided. Later that day, the United States joined a group of European countries and

Canada in announcing an additional contribution of $248 million to the Least Developed Countries Fund, and the next day the president pledged $30 million to support the expansion of three insurance funds, one for the Caribbean, one for the Pacific islands, and one for Africa, designed to provide immediate relief funds in the event of disasters.[13] These were small commitments, meant to supply some modest help and set a good tone as the conference began.

Not long after Presidents Obama and Xi made their speeches, they and their teams sat down together. The meeting was cordial, though not personally warm. Both leaders underscored the importance of climate change, their own cooperation, and getting the deal done in Paris. Xi, in words echoing the suggestion I'd made to Minister Xie all the way back in March 2009, said that climate change had become a bright spot in the US-China relationship. (The translator said "spotlight," but we got the point.) Xi also repeated his constant refrain about the "new model of major country relations."

In the midafternoon, Obama was awaiting Indian prime minister Modi, who was running a few minutes late. I was in the president's holding room when someone pointed out that Obama had wandered into the corridor and was talking to one of the Indians. I jumped up to see what was going on, and found Obama and Minister Javadekar in full debate mode, with Javadekar defending India's posture on the firewall, and Obama arguing that we couldn't solve the problem that way. Javadekar, his beard notwithstanding, looked like a kid in a candy store, flattered that the rock star president of the United States was engaging with him and delighted that he was holding his ground, refusing to give the president an inch. Brian told me afterward that he thought Obama was a little frustrated that he wasn't able to really negotiate this time—no doubt remembering his chaotic glory day in Copenhagen six years earlier. He was ready to roll up his sleeves and get at it, so when Javadekar turned up, Obama grabbed the opportunity. After five more minutes of this, Modi finally arrived.

The mood in the Modi bilat was warm and personal, and the meeting lasted well past its scheduled time. In comments at the top for the short press segment, the president called Modi "his friend," and said both he and Modi agreed climate change was an urgent threat and that "Paris gives us a chance."[14] Modi, speaking in Hindi, said, "I have such a deep relationship that we are able to openly discuss all issues" and declared that he was happy to work "shoulder to shoulder with the United States." At the end of his comments, he switched to English and remarked, "Strong message: Paris to protect planet." After the press left, and after some good-natured back and forth about growing up poor, Obama appealed to Modi to instruct his negotiators to be more flexible on key elements, including transparency. Modi pushed for increased funding, and Obama answered that adequate funds could not be raised in developed countries unless developing states were a bigger part of the solution. They also talked some about their own domestic challenges on climate change, the United States and India sharing difficult politics at home, and reaffirmed their commitment to work together.

By the time the first day ended, it had clearly been a worthwhile exercise. The presence of 150 leaders provided a vital political charge, and Obama's own meetings were quite useful. But from the negotiators and climate ministers' perspective, the day carried with it an odd sense of suspended animation since the main business of negotiating—of shaping a sizable text laden with sharply conflicting, bracketed options into concise language acceptable to all—could not get going for real until the leaders left the stage. I missed my usual habit of getting my sea legs under me through meeting with my team as well as a variety of negotiators and ministers whose judgment I trusted and valued. When I got back to our delegation suite after Obama and Secretary Kerry had left the site and the business of the day was done, I saw a message that Michael Jacobs had stopped by to see me. I first met him in 2009 in Copenhagen when he was serving as Gordon Brown's top adviser on climate change, and he had seemed a kindred spirit right away, a like-minded member of the political species who operate at the intersection of policy and politics. Small, wiry, and intense, with a big brain and the soul of an activist, some time spent with Michael felt like just the right tonic, so I called him back and we met for dinner in the simple dining area of the airport Marriott where my team and I were staying. We talked through the key issues and where the important players stood. I told him that if the main negotiators were poker players sitting around a table, I'd say that those on China's side, who wanted a minimalist agreement, had the easier hand to play because preventing something from happening is usually simpler than making something happen. And Minister Xie's deputy, Su Wei, was a world-class obstructionist when he wanted to be. Looking back, I suspect my assessment had more to do with pregame jitters than sound analysis. In any event, once the negotiating game got going, I never thought about the poker metaphor again. Indeed, by the time dinner was over, I felt grounded. Michael, as always, had been a good sounding board with his finger on the pulse of both countries and civil society—an adept listener, with well-informed views and useful advice.

* * *

The next morning, Obama met with island leaders from Papua New Guinea, Barbados, Kiribati, Saint Lucia, and the Marshall Islands.[15] This was Obama's third and last meeting, and an important part of our strategy to revive the progressive coalition after the G77 backlash at the Bonn meeting in October. Obama was great, introducing himself as "an island boy," and engaging seriously and substantively. Marshall Islands president Christopher Loeak noted how pleased the leaders were that Obama chose to meet with them as his only other meeting in Paris after China and India. They were also pleased that Obama thought of himself as an island boy. They congratulated him on his speech in the plenary, commenting on his commitment to reaching an agreement in Paris along with his recognition of the special vulnerabilities and circumstances of small island developing states. Obama stressed that all countries have to be part of the solution, targets should be ratcheted up every five years, and we needed strong transparency provisions and an effective Green Climate Fund. Then he added, "Within the G77, don't let your agenda be hijacked. That'll be the attempt, to push for a low-ambition agreement. If we allow that to happen, we'll have missed our chance." And he urged them to make common cause with vulnerable states in Africa, Latin America, and Asia to put pressure on the big emerging countries. At the end, Obama gave an impromptu summary to the press, and after the press left, had a funny exchange with President Tong of Kiribati. Tong noted that Obama had pronounced his island "kir-i-bat-i" as it's spelled, but that it's actually pronounced "kir-i-bass." Tong explained, "We don't have an *s* in our language." Obama, without missing a beat and to general laughter, replied, "We'll send you one."

The First Week

On Sunday, November 29, the day before the COP started, I met with Ashok Lavasa at his hotel in Paris, the legendary Plaza Athénée, for a one-on-one conversation—a format I requested so he would feel free to talk without having to defer to Minister Javadekar. Lavasa is a gentle, intelligent, serious man who radiates a sense of integrity. He was concerned about how Indian negotiators could defend an agreement back home if India had to give up its historical position on the firewall. I said that nationally determined targets fully protect India's ability to pursue its priorities of development and eradicating poverty, and that the United States would be crucial in securing a legal form that both our countries could accept. Although we did not make much concrete progress, I thought the meeting was useful because we had a candid and friendly exchange that I hoped would set the stage for more constructive dialogue.

* * *

I met with Minister Xie nine times in the eleven days of Paris (not counting the meeting between our presidents that we both attended). This in itself was an important indicator that whatever disagreements we had on issues, we were aligned on the proposition that the United States and China had a special relationship, special capacity to influence the outcome of the conference, and common commitment to deliver on the promise of our leaders' 2014 Joint Announcement. That said, we did not make progress in the first week. We talked, joked a little (Xie claimed he told US media that our relationship was so close, "Stern's wife is a bit jealous"), and sparred a lot, but did not get very far.

Our meetings, on December 1, 2, 3, and 6, returned repeatedly to Xie's assertion that he was assiduously defending the United States against the many parties that opposed our position on the legal form of the agreement—by which he meant nonlegally binding emission targets but legally binding provisions for process elements such as transparency, the need to submit new targets periodically, and so on. At the first of these meetings, he said the majority of countries insisted on a fully legally binding agreement, and that he was "concerned" that this could put the United States in a difficult position. On December 2, he explained he was defending our legal position against criticism from progressive Latins. On December 3, he said a developing country delegate asked why transparency should be legally binding if emission reduction targets were not, and that when he asked this person whether he wanted a repeat of the Kyoto Protocol with the United States outside the agreement, the person answered, "But why not consider our preferences. We don't like legally binding transparency." At this point, the amused thought that crossed my mind was that Xie had an "imaginary" friend like our eldest son had when he was two or three years old and used to tell my wife and me all about his friends Parch, Larch, and Dondon. In any event, it was now clear what this daily narrative was all about: it had been the windup and now came the pitch: "So we have to work on those countries (in regard to transparency), push them to accept what we said in our joint statement from September, and even this would not be easy." In other words, we should pocket the five lines on transparency from our September joint statement and not expect anything more. Minimalism in action.

On Sunday, December 6, Xie picked up the thread again. He said that the European Union, New Zealand, developing countries, UNFCCC executive secretary Christiana Figueres, and even Ban Ki-moon had asked him about the position of the United States, and that he told them he fully understood the position of Todd Stern and also pointed out that Obama was doing more on climate change than any previous US president. He said he told everyone he talked to that they needed to show flexibility to the United States, but that others said that if they do that, the United States should show flexibility to them. Once again, a full-scale windup, but this time a different pitch. This time he argued that the overarching Article 2 of the agreement should say that the agreement's purpose was "to enhance the implementation of the [Framework] Convention." He noted, "I hope you can show flexibility on this." But I could not show flexibility. If the Paris Agreement described its purpose as implementing the Framework Convention, China and its LMDC colleagues would immediately read that language to include the Annex 1 / non-Annex 1 structure as well as other principles and provisions routinely claimed as embodying the firewall. They would then invoke the language about the purpose of the Paris Agreement to justify bifurcation in implementing the agreement and drafting detailed guidelines in the coming years.

* * *

On Friday night, December 4, I had dinner with Luiz Figueiredo and his colleague Rafael Azeredo at La Maison de l'Aubrac, a steakhouse in Paris. The atmosphere at dinner was relaxed and cordial. Coming off our successful joint efforts preparing President Rousseff's visit to Washington in June, Luiz and I shared a kind of unspoken expectation that we would work together in Paris toward a result in the interests of both countries. That Luiz was the current Brazilian ambassador to the United States, with a natural interest in constructive relations between the two countries, didn't hurt. We had a productive exchange about differentiation. He agreed that the nationally determined model for NDCs made sense rather than bifurcation and thought it could be applied to transparency as well. This was not literally true since the basic transparency rules had to be agreed on in a top-down manner; it would not work for every country to design its own system of reporting and review. But the idea of flexibility, as we incorporated it in our joint statement with China, would indeed provide an element of national determination. I came away from dinner thinking we had made good progress.

* * *

On Friday morning, I met with Tony de Brum and Jimmy Fletcher. They saw 1.5°C as a critical threshold issue for the islands, and were looking to the United States to show flexibility on loss and damage. In the course of our conversation, we talked about how some countries were pushing for a minimalist outcome while many, including the islands and other progressives, were pushing hard for high ambition. Tony referred to this group as a "high-ambition coalition." I liked the phrase and thought it could be useful. At my public press briefing later that day, I said I was "encouraged by what might be called a high-ambition coalition emerging here. It includes many countries; it does not include everyone. It is clear that a great many countries want an ambitious agreement, and we are certainly in that camp." In using Tony's phrase, I meant to start defining a new dividing line in the negotiations, high ambition versus low rather than developing countries versus developed, to energize those pushing for ambition and put those who were resisting on notice that we would call them out.

I again beat the drum on what became the High Ambition Coalition (HAC) at my briefing on Monday, December 7, explaining that it included the island states, LDCs, developing country progressives, Europe, and members of the Umbrella Group, like us. I listed the central elements of high ambition, including the five-year cycles. I embraced the notion of a reference to 1.5°C. I outlined the elements we wanted to see in a good transparency article—reporting and review of inventories and progress in implementing NDCs—with enough detail to make these basic elements clear. I said I imagined the transparency article would be a page or two, but it could not be a sentence or two, mindful that the transparency paragraph in the US-China joint statement from September was two sentences comprising five short lines. On Sunday night, December 6, Tony, together with the United Kingdom, hosted a HAC dinner for roughly twenty-five people at Drouant, a restaurant in Paris. The dinner was long and the discussion a bit rambling, but that was fine. Tony'sobjective was to build relationships and solidarity, and the meal did both of those things.

The Second Week

Secretary Kerry, who attended the first day of the COP with President Obama, returned on Monday, December 7, and stayed until the end, to good effect. He, Sue, and I met with Xie on Tuesday afternoon and focused on transparency. Xie returned to his December 3 position that we already had our solution—the five lines he and I along with Sue and Su Wei negotiated for the Obama-Xi joint statement in September. I was surprised to hear him still arguing this with only three and a half days left in the negotiations. A full transparency article required language about the operational elements of reporting and review, among other things, and our joint statement language did not do that. Moreover, using only limited lines would omit the language and options that Fook Seng had passed along to the French from the transparency group he had been facilitating. Kerry said that the joint statement language was not enough, referring to it as a "press statement," which visibly offended Xie, who retorted that "China was carrying work out in a serious manner, not just a press statement." Kerry had to leave for another commitment not long after this minor contretemps, and I picked up the thread. The joint statement language on transparency was important, especially the sentence on flexibility, but it was bare bones: All it said was that

> both sides support the inclusion in the Paris outcome of an enhanced transparency system to build mutual trust and confidence and promote effective implementation including through reporting and review of action and support in an appropriate manner. It should provide flexibility to those developing countries that need it in light of their capacities.

Xie's position again underscored China's hope to keep the Paris text at the level of high principles, so that the operational paragraphs could be hashed out later outside the intensive scrutiny and political pressure of Paris, in a venue where it would be easier to delay progress toward a unified system. By the end of our session, Xie acknowledged that we would not accept China's limited preference and suggested that we give him any additional language we thought necessary. This constituted some forward motion, though I walked out of the meeting annoyed because I thought it time to get past this kind of posturing. I decided to avoid Xie over the next day or two to convey the message that I did not intend to continue in this manner. It was time to get down to business.

* * *

We met frequently with the Indians through the two weeks of the conference. On December 8, I joined Deese and Kerry for a meeting with Javadekar, Lavasa, Dr. J. M. Mauskar, and others, but we again got nowhere on the core question of one transparency system or two. Kerry tried to woo them with an odd promise from Solar City (whose board chair was Elon Musk) to make some kind of special deal for developing countries that wanted solar power—Kerry described it loosely as "free solar"—which both intrigued and confused the Indians, but did not affect our discussions. Sue and Paul Bodnar met again later in the week with Lavasa, and Sue suggested that they both put aside the current text and just identify what each of us wanted to be able to say at home. This was helpful, and she jotted down the various comments they both made and fed the ideas into the French transparency process.

* * *

The next day, Wednesday, December 9, proved to be the pivotal day for COP 21. It began for me with a HAC brunch hosted by the Marshall Islands and United Kingdom in the UK's delegation space. Some thirty to forty people crowded around a long, thin oval table. There was some talk about how the HAC was gaining traction. The *Guardian* had run a feature story the day before, with the headline "Climate Coalition Breaks Cover in Paris to Push for Binding and Ambitious Deal."[16] During the brunch, we agreed that a small group of us would hold a press conference later that day as a kind of official coming out party, led by Tony.

The press conference took place at around 7:00 p.m. in a packed briefing room. The tableau itself told a story: de Brum presided in the middle with

Mexico, Colombia, Angola, and the European Union (Cañete) to his right, and Germany, Gambia, the United States, and Norway to his left. Earlier that day, *Climate Home News* had said on its blog, "Rich and poor countries are catching the likes of China and India in a 'high ambition' pincer movement."[17] The new alliance was there for all to see—a long way from Diseko silencing progressive voices through G77 "discipline" at the October session in Bonn.

Tony started off, saying that a high ambition coalition was emerging at the conference made up of more than ninety developed and developing countries. He said it was not a negotiating group but instead "about joining the voices of all those who are committed to an ambitious agreement and a safe climate future. Big and small, rich and poor." He noted that we would not accept a minimalist agreement and would be fighting for basic objectives, including an "ambition mechanism, which I consider the beating heart of the Paris agreement," along with recognition of the below 1.5°C temperature goal, a clear pathway for a low-carbon future, and five-year updates of NDCs. Everyone spoke briefly, and Tony called on me first, recognizing that full-throated US participation in this new progressive coalition was news. I reiterated the objectives Tony had named, adding a strong transparency article so everyone would know how others were implementing their NDCs. And I said, "The High Ambition Coalition is exactly what we need right now. There are a great many countries in this coalition; there are some countries that are not in this coalition and indeed that seek a more minimal outcome." The first question from reporters asked me to name the countries I was alluding to. I demurred.

But the truth was that India and China, among others, had reservations, especially about the five-year cycles. The point of the individual cycle was to require countries to review and urge them to ratchet up their targets to higher levels every five years.[18] The point of the staggered five-year global stocktake was to review how all countries in the aggregate were doing in meeting the climate challenge, taking account of the latest science. India argued that five-year cycles should be a "matter of choice." China appeared not to want five-year cycles at all. Neither wanted provisions that opened them up to being judged, criticized, or pressured to do more. But Tony and Cañete were both adamant. Tony said that anyone trying to water down the cycles was "not committed to urgent climate action."[19] Cañete said that "without five-year cycles, the agreement is meaningless."[20]

* * *

Secretary Kerry and I did three meetings together on December 9—with Vivian Balakrishan of Singapore, Brazil's Luiz Figueiredo, both facilitators on differentiation, and Laurent Fabius. With Vivian, Kerry homed in on a transparency paragraph of the new December 9 draft text, the first issued by France since it took over the text from the cochairs of the negotiation on December 5. The problem was, first, that the first of three options could be read to imply bifurcated transparency. I told Kerry before these meetings that we needed to make it clear that there was a point beyond which the United States would say no, even if it meant preventing the agreement from going forward. I thought it important to puncture any impression that the United States would never walk away from a Paris deal given how consequential the conference was. And I thought it would have added impact if Kerry made the point to Vivian and Luiz. Kerry said to Vivian, "Despite what is said about how this is the moment, we won't do an agreement in the old [bifurcated] manner. We are not going to do a suicide pact." And he made the same point in our meeting with Luiz. Before that meeting ended, Luiz, never forgetting his day job as Brazil's ambassador to the United States, asked Kerry if he would come to Brazil next year. Kerry, never forgetting the art of dealmaking, said yes.

Sue joined Kerry and me for a meeting that same day with Fabius, a colleague of Kerry's in the exclusive club of foreign ministers and someone he had known for years. Kerry pushed for a unified transparency system as crucial to the credibility of an agreement that would be self-differentiated and nonpunitive. He also told Fabius that we needed language making it clear that including loss and damage in the agreement would not provide a basis for claims of liability and compensation against developed countries.

* * *

At nearly midnight Wednesday, I was on my way to the large room where the first Indaba meeting was going to start soon, with roughly eighty participants sitting around an enormous rectangular table to discuss their views about the new December 9 text.[21] Minister Xie saw me and eagerly motioned me to follow him into a small room on the other side of the wide corridor. I could tell from his body language that he was in a completely different frame of mind than when I had last seen him a day and a half earlier. Now he looked ready to do business. Brian was with me, and we both followed Xie into the room, where he was joined by Su Wei. In short order, Xie said, "Now is the time for the United States and China to play our role and find a solution." He wanted to talk about transparency, mitigation, and finance. He said that he thought all of those issues required political decisions, and that if we could find common ground, we could give our joint proposals to Fabius. I agreed, but said we also needed to talk about the proposed "purpose" language in Article 2 and another up-front, overarching article ("2bis") that we saw as loading the dice in favor of bifurcation. On transparency, I agreed that the language from our September joint statement should be the starting point, but said we also needed language on how the basic functions of reporting and review would work. I said that the current draft had useful language on these points. Xie agreed as long as we preserved our joint statement language.

The mitigation issue he was concerned about involved the sequencing of the five-year cycles. He was still focused, as he had been in Lima, on beating back any hint of an assessment of proposed new NDCs, which he feared would happen if the global stocktake was held right after countries announced their new NDCs—a sequence advocated by South Africa, among others.[22] Xie wanted the global stocktake to occur two or three years before the individual country reviews to ramp up NDCs—to influence those NDCs before the fact rather than passing judgment on them after they were announced.

I thought that was fine and did not see South Africa's proposal getting traction in any event, but I told Xie that I needed one thing from him in return: in regard to five-year country updates, whether a country decided to ratchet up its target or leave it alone, it would need to say so in writing and submit that writing to the UNFCCC secretariat explaining what it had decided. China and India had opposed this idea, arguing that if a country decided simply to maintain its existing target, it should not need to communicate anything. But I felt strongly that insisting on a communication would put useful pressure on countries to ratchet up after all, and if not, at least be forced to state that publicly. So I insisted that silence was not enough. Having secured the point he wanted on the sequencing of the global stocktake and new NDCs, Xie agreed that countries would have to submit the kind of communication I wanted.

Regarding Articles 2 and 2bis, I told Xie that the parties were trying to put a sharply bifurcated overlay on the whole agreement, which would destroy the carefully balanced differentiation that we were negotiating in areas such as transparency. He didn't push back. We simply agreed that we would once again enlist the two Sues, Sue B and Su Wei, to take the basic guidance we had agreed on in this meeting and turn it into suggested text.

* * *

Sue, Brian, and I met again with Xie and his team the next morning, December 10, at 10:15 a.m. Xie and I talked briefly, with the only new point being Xie's suggestion that developing countries be given "incentives" in the form of financial or technology support to encourage them to ramp up their NDCs. I said we could not agree to that, and we moved on. The two Su(e)s started their work at around 11:00 a.m. and finished seven hours later. They made good progress.

Their version of the introductory Article 2 no longer included language saying that the purpose of the agreement was to "enhance the implementation of the Convention." It said that the agreement, "in enhancing the implementation of the Convention, including its objective, aims to strengthen the global response to the threat of climate change." This small change made a big difference. Now, the purpose of the agreement was to strengthen the global response to climate change. "Enhancing the implementation of the Convention" was retained, but as a subordinate clause, secondary to the main purpose. No one could rationally argue that this subordinate clause meant that every provision in the Framework Convention was effectively retained. The two Su(e)s also negotiated language for what became Article 4.4 of the final agreement in a manner that softened the way that differentiation was expressed, though this short paragraph was far from out of the woods, as we shall see.[23] And they agreed to incorporate broader language on the operational aspects of transparency that had been negotiated in the group led by Fook Seng.

Xie and I went together to meet with Fabius in the early evening to explain the work we had done together. Fabius took the short paper on which Sue had recorded what she and Su Wei had agreed, and passed it along to Anne-Sophie Cerisola, one of the talented members of Laurence Tubiana's team, recognizing it as containing important political solutions. He instructed her to have the language included in the text. The next crucial step, from our perspective, was for Xie to bring along the members of the LMDC.

* * *

The French issued a new text at around 9:00 p.m. on Thursday, December 10, and then held another Indaba, starting around midnight, to hear reactions and concerns from the parties. Secretary Kerry attended, and when it was his turn to speak, he lectured his audience. It was time to stop all the petty arguments, he said; we were facing an enormous threat and the text was already a "monument to differentiation" because of its nationally determined structure. He told everyone to stop fussing and quibbling, and just get the deal done. I wasn't sure how all of this was going to go over, and when I saw delegates from LMDC countries like Sudan, Argentina, and Malaysia headed for me at the end of the meeting, I assumed the worst—protests over Kerry vigorously waving away the idea of an ongoing firewall, almost scolding the room. But I was wrong. They praised his intervention—probably for a few reasons. Kerry was an impressive figure, tall and stentorian, a polished, confident speaker, the US secretary of state who had given an urgent speech about climate change that same afternoon and then, after midnight, come to a venue that didn't usually see people of his stature sitting around a big rectangular table with climate negotiators and ministers. Plus when he called them out about the squabbling pettiness, many realized he was speaking the truth. Which didn't mean they would suddenly support our positions, but it did mean that his performance and message had broken through.

* * *

My unexpected pleasure on Thursday was running into my old Danish friend Bo in the late afternoon as I headed to a plenary session. I had a few minutes to spare, so grabbed Bo to sit with me while I ate a piece of bad pizza. He was there to cover the conference for the Danish newspaper *Politiken*, where he had become editor in chief in 2011. Our chance encounter seemed perfectly timed, as if the hint of a circle that was about to close.

* * *

Negotiations continued for most of the night on Thursday. When I returned to Le Bourget Friday morning, Brian and I did a quick check-in meeting with Xie, where we learned that President Obama had talked to President Xi. Earlier in the week, Obama had already called Hollande, Modi, and Rousseff, putting his shoulder to the wheel one more time. Xie also told me he met with Malaysia's fiery Gurdial Singh Nijar, the LMDC representative who had temporarily stolen the show in Lima, and stressed that in this final stage, we all needed to show flexibility. He was also keen on Fabius using the text of what the United States and China had agreed to Wednesday night and Thursday. I told him Kerry and I were going to meet with Fabius soon to see where we stood. Xie said, "We are still one step away, so let's work hard. Tell Kerry if we get agreement, he promised to treat me to dinner."

* * *

The big news out of the HAC on Friday was that Brazil had joined, another jolt of momentum. *Climate Home News* reported, "It's official: Brazil is part of the high ambition gang, splitting from the rest of the BASIC bloc: China, India and South Africa."[24] Brazil's Teixeira said in a written statement, "If you want to tackle climate change you need ambition and political will. . . . Brazil proudly supports the high ambition coalition and pledges our political support to this effort."[25] The HAC, according to *Climate Home News*, had "burst onto the scene this week in Paris, back[ing] a five-year review mechanism, a long-term goal to cut emissions, and a recognition of the dangers of warming over 1.5C."[26] Cañete tweeted, "Game changer: Brazil joins the EU, US, 100+ developing countries in the #HighAmbitionCoalition!"[27]

* * *

The December 10 text, however, had produced one genuine problem. While it made some useful progress on ambition and differentiation as compared to the December 9 text, on finance it took a sharp turn for the worse from our perspective. It declared in unbracketed language that the amount of mobilized funding in the post-2020 period "shall represent a progression" from a floor of $100 billion per year, and that "short-term collective quantified goals" would be "periodically established and reviewed" by the parties to the new agreement during the five-year global stocktakes. But donor countries had not agreed on a rising progression of funds beyond $100 billion post-2020 nor to periodic new short-term quantified finance goals. This shift in the text toward our side on ambition and differentiation, but toward the developing country side on finance, was a deliberate tactical move by the French, as Rémy Rioux, a senior member of the French team, subsequently explained in a 2018 report about COP 21.[28] By pleasing developing countries on finance, the French meant to give themselves room in the December 10 draft to do what they thought necessary on ambition and differentiation. In addition, they wanted to demonstrate to developing country negotiators, including South Africa's Diseko in her role as chair of the G77, that France was prepared to move in their direction on finance assuming the move had adequate support among the parties. If the new language failed as a result of too much donor country opposition, France would at least have done its best.[29]

The French made another important move on finance in the last days of the conference as a matter of substance more than tactics. With Laurence in the lead, France secured language in Article 2, the purpose article, articulating the imperative of making all public and private financial flows going forward "consistent with a pathway toward low greenhouse gas emissions and climate resilient development."

* * *

The effort to land both the global temperature and accompanying emissions goals went down to the wire. The aim of holding the global temperature increase to 1.5°C or less was a passionate priority for the island states and LDCs. The United States and European Union favored making the traditional below 2°C the main goal, while supporting a reference to 1.5°C as well. During the final Indaba session, at around 3:00 a.m. Thursday night, Minister Fabius asked various facilitators to convene breakout sessions, one of which was on the global temperature goal, led by Norway's Tine Sundhoft and Saint Lucia's Jimmy Fletcher. More than sixty negotiators attended. Eventually, the language that ended up in Article 2.1.a was proposed—holding average temperature to "well below 2°C above preindustrial levels and pursuing efforts to limit the temperature increase to 1.5°C . . . recognizing that this would significantly reduce the risks and impacts of climate change." Trigg was there for our team, and Pete Betts (United Kingdom and European Union) asked him if the United States could accept that language. Trigg said, "We can sort of probably buy it," at which point Pete concluded, "It's done if the Yanks can buy it," and it was.

China and India were not happy, but accepted the 1.5 language because it was a best-efforts goal.

The French were determined to establish an emission reduction goal too—something the influential climate activist and adviser (to the Marshall Islands in Paris) Farhana Yamin had been pushing hard for years under the banner of "net zero."[30] Farhana's sound logic was that people can't get a handle on what holding a temperature increase to 2 degrees means in terms of ordinary activities, but they know what producing and reducing emissions means and they know what zero means and the views of the public are enormously important if we hope to decarbonize the global economy.[31] Laurence later said that a variety of formulas had been proposed and rejected—net zero, zero greenhouse gases, carbon neutrality, climate neutrality, and greenhouse gas emissions neutrality, among others—all of which were rejected. On the last day, the French proposed more technical-sounding language about "achieving a balance between anthropogenic emissions by sources and removals by sinks"—in other words, a balance between greenhouse gases released and absorbed. This formula worked.

* * *

Sue had a busy day Friday, negotiating final elements of text in three different areas. On loss and damage, the French told us we needed to find common ground not just with the likes of the Marshall Islands and Saint Lucia but also Tuvalu, the most unyielding of the island group. Secretary Kerry and Sue had already met with Prime Minister Sopoaga on Monday, but it was time for another round, and Sue proposed that Kerry meet with Sopoaga alone, leaving her and her hardheaded counterpart, Ian Fry, outside the room. She didn't want the potential for understanding between Kerry and the prime minister to get sidetracked by protests from Fry, who was as pugnacious as they come. Sue and Christina Chan from our team had been working constructively with the small islands in the last days of the conference on the broad loss and damage article, but it was the matter of liability and compensation that required the final meeting with Tuvalu. It turned out that Sopoaga had not fully processed what Sue and I had told de Brum and Fletcher at the pre-COP meeting in November: that while we needed language making it clear that the agreement does not provide a basis for liability or compensation, we never meant to strip away whatever rights developing countries might be deemed by a court to have already. When Sopoaga heard that reassurance from Kerry as well as our acceptance of Tuvalu's request to put the liability and compensation disclaimer in the accompanying decision rather than in the agreement itself, the loss and damage deal was done.

Sue also worked on the nuanced language of what became Article 4.2, giving as much legal force to mitigation efforts as we could, consistent with our own redline. For weeks, the European Union and its progressive allies had been agitating for words that sounded as legally binding as possible without crossing our trip wire. At the pre-COP, for example, having recognized that the United States could not accept "shall achieve" their target, the European Union and its allies pushed for "shall implement," but we could not accept that either. Sue's work on Article 4.2 was deft, as usual. In the final version, the first sentence stated the basic legally binding obligation of every party to prepare, communicate, and maintain successive NDCs. The second sentence has two parts, the first of which says that "parties shall pursue domestic mitigation measures"—a legally binding commitment that we were comfortable accepting since it did not have a specific target attached to it and was akin to a commitment we had already made in the Framework Convention. The second part says "*with the aim of* achieving the objectives of such contributions"—a nonlegally binding formulation about the targets themselves.

Finally, Sue and Paul Bodnar worked on nailing down language for the finance article with France's Rémy Rioux and South Africa's Sandea De Wet, a tough, skilled lawyer representing the G77. The final draft included a legally binding obligation for developed countries to provide financial resources, but one that was "in continuation of their existing obligations under the Convention" and therefore would not trigger an obligation to secure Senate approval of the new agreement. The text included modest encouragement to wealthy enough developing countries to provide support voluntarily and "as part of a global effort." Developed countries continued their existing pledge to mobilize $100 billion per year through 2025 (from all sources, and in the context of meaningful mitigation actions and transparency), and for the period after 2025, the parties to the Paris Agreement would "set a new collective quantified goal from a floor of USD 100 billion." The language about maintaining the existing pledge through 2025 and setting a new goal of at least $100 billion before 2025 was critical for developing countries. There would have been no Paris Agreement without it. At the same time, leaving open the question of who would be expected to contribute post-2025 and the amount of funding, other than to say it would not shrink, was crucial for developed countries.

* * *

By Friday evening, other than an occasional last bit of business, an odd period of calm settled over the conference center—except of course, for the French team members, who were still killing themselves to get a final draft prepared by 9:00 a.m. the next morning, as Fabius had promised. In the last twenty-four hours or so, Laurence Tubiana and her team looked haggard, on the far side of exhaustion from too many sleepless nights and too much pressure. Plus as the COP presidency, they carried the additional and inescapable burden in a climate conference that you just never know. Even if you think you're OK, even if all the pieces have seemed to come together, you know there will be some discontented parties, and if even a small group of them decide in the end that they cannot swallow the final text, the whole conference can come undone. At least two members of the French team—Laurence and Emmanuel Guérin—were eyewitnesses to the crash in the Copenhagen plenary hall in 2009 and were still scarred by that memory. Emmanuel later described his exhausted state of mind as the last Paris plenary began in one word: fear. The French team did not rest easy until the final gavel came down. And as events played out, the team was right to stay on high alert.

As for me, I was restless, and at around 9:00 or 10:00 p.m. started wandering. Downtime was nothing I was used to and nothing I liked during negotiations. I visited the UK suite and talked to Minister Amber Rudd and Pete Betts. I visited Andrea Guerrero in the Colombia delegation. I ate something at one of the cafeterias with Paul. I passed by South African Minister Edna Molewa in the corridor, then the Africa group, and then LMDC leaders, pausing to joke around with Malaysia's firebrand Gurdial Singh Nijar.[32] That negotiators can go tooth and nail at each other in the heat of battle and then, not long after, share a friendly moment might seem strange from the outside looking in, even inappropriate, as if to suggest that the whole argument was just a charade. But it is a quality of the negotiating life that I always liked. The debates and disagreements are certainly real, but so is the recognition by those of us in the thick of it that we were all working for our countries, all enmeshed in a tense, high-stakes, difficult game, and in most cases, were people worthy of respect and a touch of friendship.

I walked on toward the offices of the French presidency and saw Angola's Giza Gaspar-Martins on the way. Then I paid a short visit to Fabius, just one-on-one. I had a couple things to say, but mostly wanted to just make a human connection. Here we were at the end. The French had given it everything they had, the conference hall was quiet, and the negotiating mostly done. I found Fabius alone, and he welcomed me to sit down with him. He too looked worn. My first piece of business was to ask why he was so keen on referencing the $100 billion in the agreement rather than the accompanying decision. When Secretary Kerry and I had talked with him earlier that day, he made it seem like this was essential for developing countries, but I had done a little sleuthing of my own, including with Andrea and Giza, the LDC lead, and I wasn't detecting much heat regarding the textual placement of the $100 billion. I relayed what I knew to Fabius and told him that we, and I suspected others, would prefer the number in the decision where it would have no less force, but would call less attention to itself. He listened but did not answer. I also told him, in friendship more than anything else, that he shouldn't put too much pressure on himself or his team to get the final draft out exactly at 9:00 a.m. If it came out at 11:00 a.m. or 1:00 p.m., that would be fine, no one would care, and it was much more important to get it right than to worry about a few hours here or there. Fabius was typically a stickler about time, but I think he appreciated my advice and visit. After that, I stopped by Singapore's suite and talked briefly with Vivian Balakrishnan, "Minister Vivian" as those on the Singapore delegation called him, and Fook Seng. Not long after midnight, I went back to our hotel with some of our team.

The Final Day

Saturday, December 12, was a remarkable day. I arrived at Le Bourget around 9:00 a.m. and went with Brian, Sue, and Paul to meet Xie and his team to do a signals check that everything seemed on track. After 11:00 a.m., I went to the United Kingdom's delegation space to join a large gathering of the HAC, and at around 11:40 a.m., we started to march toward La Seine plenary hall with Tony de Brum, Cañete, and me in the front, the group walking elbow to elbow, sometimes arm in arm, with people in the crowded corridors cheering along the way, television and still cameras everywhere. As we marched into the plenary, the hall burst into applause. Much later, France's Guérin told me he thought the image of the United States walking in with the small islands—Tony and I side by side, an unexpected image—"won it," making it hard for naysayers to stand up in opposition. Fabius and President Hollande each delivered strong speeches, and as I sat next to Secretary Kerry behind the US placard, with Sue and Trigg behind me, and nothing left to do, the realization sunk in that this agreement was actually going to get done. That nothing would stop it now. That all of our effort and all the efforts of so many people around the world for so long, inside and outside governments, were finally going to succeed. I jotted down an email to my wife, Jen: "Almost there. Amazing, really. Could still hang up, but there's a tremendous sense in the room that it will happen. Great speech by Laurent Fabius. An inspiring march by the HAC. We'll see. Living history."

What did not strike me as odd at the time, swept up in the excitement and emotion of the moment, was that here we were giving Fabius and Hollande a standing ovation, but *no one outside of the French team and secretariat had seen a text in nearly two days*. By COP standards, this was nothing short of extraordinary. Negotiators typically hang on the final draft text that emerges on the last day, read it intently, consult within their country teams, and meet with their groups. The Fabius and Hollande speeches were, in retrospect, an astonishing bit of French stagecraft. No doubt the French did feel confident that they had the political balance right in the final text and also that most parties felt that they had been heard. And yet it was still certainly possible that some small group of countries would find the final December 12 text unacceptable owing to language it contained or omitted. So the stagecraft was also a gamble. The French made the bet that by creating a sense of inevitability, a sense that the deal was done, they would minimize the chance that any country or group would have the nerve to stand up and say no. But there was still the risk: of a simple but fateful mistake that might be difficult to repair.

* * *

The final draft came out around 1:30 p.m. I grabbed the first copy that came off our printer and sat down alone in our large conference room. When I reached Article 4.4, I froze. It said "developed countries *shall* continue taking the lead by undertaking economy-wide absolute emission reduction targets" where it was supposed to say *should* continue, and "shall" is a telltale signifier of legal bindingness in international agreements. Thus the mortal danger here was that this one word change, creating a legally binding commitment to adopt a certain kind of target, would force us to submit the agreement to the Senate, where securing the constitutionally required sixty-seven votes was out of the question.[33]

The change was simply inexplicable. An earlier version of the same paragraph in the December 10 text had said "should." When Minister Xie and I met with Fabius on Thursday evening to show him the language suggestions the United States and China had agreed on, that paragraph was included because of related changes Sue and Su Wei had agreed on, and that new version of the paragraph also said "should." I called Sue to come take a look, and she was just as startled and worried as I was. If the final agreement included "shall," either we would have to submit it for approval to the Senate or face a firestorm of opposition for not doing so. When Kerry walked in to look—a man of the Senate for twenty-eight years—his reaction was emphatic, and he cut off any reflections about whether we could still somehow argue that Senate approval was unnecessary. He too thought the language would kill us. He called Fabius, who said he had no idea what happened. Sue called the French team's Anne-Sophie Cerisola, who said the same thing. We told the French that we could not accept the agreement unless the mistake was fixed.

Who changed "should" to "shall" was a mystery, though one we had no time to focus on in the moment. It did not happen by autocorrect nor was it a typo. Someone did it deliberately, probably thinking they could make NDCs sound a little more legally binding while not crossing the US tripwire, unaware of the nuances of US treaty law—a dreadful miscalculation. The only people with access to the documents were in the UNFCCC secretariat and the French delegation, though in the aftermath I've never heard anyone who believed a person on the French team did it, sensitive as they were to the US obsession about legal form. What is certain is that on any informed cost-benefit analysis—the minor upside of making the proponents of legally binding targets feel a little bit happier versus the calamitous downside potential of capsizing the entire agreement by making it impossible for the United States to join—it was a crazy gamble to take.

On the plenary floor, the risk to the agreement was quite real. In my gut, I couldn't believe that this whole, world-changing accord would unravel over something so witless, especially given the enormous momentum that had built up. The *Climate Home News* blog at 6:14 p.m. captured the mood: "It's a heady atmosphere in Le Bourget, with sporadic bursts of cheering. They're in no hurry to start, with everyone chatting and joking. At this stage, it is almost inconceivable the deal won't be approved." Yet there was the real danger that countries would believe the United States was unfairly demanding a new change, not simply insisting that an error be fixed, and the still more serious threat that a few unhappy countries would seize the opportunity to demand their own changes as the price for permitting this allegedly new US maneuver. If someone started pulling on the thread like that, there was no telling whether the whole garment would unravel. As the effort to fix the mistake dragged on, *Climate Home News*, at 7:11 p.m., caught the shifting mood: "Now, the US is arguing 'shall' was never supposed to be in there. . . . That appears to be backed up by Thursday's draft, which only used 'should.' But now it is out there, some are framing it as the US trying to weaken its commitments."[34]

I talked to Xie, who knew what we had agreed on, but was reluctant to say anything suggesting that the United States and China had worked together on language. I talked to the Saudis (Khalid Abulief), South Africa (Minister Molewa and Nozipho Diseko), and the Indians, all of whom listened to me politely, but gave no sign of what they were thinking. I talked to Amina Mohammed, then the Nigerian minister of the environment, but she was angry and aggravated for a different reason.[35] Africans had tried repeatedly to have the "special circumstances and vulnerabilities" of Africa singled out in the preamble, akin to the special recognition accorded to the LDCs. (Most of the forty-eight LDCs were African, but some important African countries were not LDCs.) At one point, near the end of the conference, she warned the French that Nigeria and other African countries might block the final text altogether unless this change was made, but Fabius told her he could not do that because the Latins were dead set against it.

Meanwhile, Fabius was mostly in a back room talking to delegations. It turned out that the Nicaraguan lead negotiator, Paul Oquist, was holding things up, partly over the mistake, and partly as a means of protesting other aspects of the deal he didn't like. Paul, who sadly died in April 2021, possibly of COVID, was a famous character around the negotiations—a large man of wide girth and oversize suits with a walrus mustache and the capacity to hold forth on most any subject at inordinate length.[36] He studied at the University of California at Los Angeles and Berkley back in the heyday of the 1960s, went to Nicaragua to work for the Sandinistas, and never came back. He was an entertaining guy to sit next to at dinner, explaining such oddities as the (once alleged) relationship between Finnish and Korean, but he could not be easily moved once he got an idea fixed in his head. Various efforts were made to reach senior government officials in Nicaragua to give Oquist instructions and reach others who might be able to persuade the Nicaraguan leadership to give him such instructions. Even the pope was enlisted to call President Daniel Ortega, though it wasn't clear whether it did any good.[37] Eventually Oquist let it be known that he would refuse to indicate support for the agreement, but would not block it either.[38]

At that point, some ninety minutes after the session was supposed to begin, the COP leadership reassembled on the dais. Richard Kinley, the deputy executive secretary of the UNFCCC, quickly read through a series of minor corrections in the text in his best deadpan, UN bureaucrat's voice, involving dropped words, misplaced commas, and the like. There were some ten or twelve of these errata, and in the middle, he unobtrusively inserted the Article 4.4 fix, not saying that "shall" was being changed to "should," but just that the opening clause of Article 4.4 should read "developed country parties should take the lead" and then continuing briskly on to the next five or six corrections. When Kinley was done, Minister Fabius said quickly, with his voice rising to meet the moment, "I am looking at the room, I see that the reaction is positive, I don't hear any objections, the Paris Agreement for the climate is accepted," and before anyone knew what had happened, he banged down his small green gavel, the floor erupted, and it was over.

* * *

On the dais, Fabius and Hollande, together with Ban Ki-moon, Laurence Tubiana, and Christiana Figueres, the UNFCCC executive secretary who had done so much to build momentum and support for the agreement among countries, civil society, and the private sector, joined hands and lifted their arms in triumph. The hall was elated—"electric with emotion" in the words of *Le Monde*, the prominent French newspaper. Fabius himself, the classic Cartesian and not usually emotive, choked up. Hollande, in his passionate speech, just four weeks after the terrorist attack that had devastated

Paris, said that "the 12th of December 2015, will remain a great date for the planet. In Paris, many revolutions have unfolded, but today, this is the most beautiful revolution, the most peaceful, the revolution for the climate."[39] *Climate Home News* described de Brum as "visibly emotional," saying, "I can go back home to my people and say we now have a pathway to survival." In Washington, President Obama called the Agreement historic and said it "represents the best chance we have to save the one planet we've got." Alf Wills, the ever-calm, ever-wise, understated, longtime South African negotiator, known everywhere for his gray ponytail, said simply, "It's a system shifter."[40]

Postmortem

And indeed it was. A paradigm shifter, better than the best-case scenarios we imagined. It was universal, its obligations and expectations applying to all countries. It was durable, built to continually renew itself. It developed a blended structure grounded on nonbinding bottom-up NDCs, the sine qua non of the Paris Agreement, and top-down rules providing assurance that countries would submit and regularly update their NDCs; that they would do this months before a COP, so the NDCs would be exposed to the sunlight; that the NDCs would include the information necessary to make them clear and understandable; that country action would be subject to a credible, robust transparency process of reporting and review; and that these basic obligations would be backed up by a committee to promote implementation and compliance. It included what Tony de Brum called the beating heart of the agreement: an ambition mechanism of long-term goals to limit temperature increase and global emissions; an initial set of NDCs from nearly all nations, not strong enough, but an important start; twin, five-year cycles, one for updating individual targets, the other for a global stocktake; an agreement that all successive NDCs would "represent a progression" beyond the targets they replaced; and a call for countries to submit midcentury low greenhouse gas emission strategies by 2020.

The Paris Agreement also incorporated a new, more supple form of differentiation, leaving the old firewall behind. This modified form of differentiation is visible in the self-differentiated structure of country targets; the transparency article's capacity-based flexibility; the new phrase "in light of different national circumstances," added to the classic "common but differentiated" formula, suggesting that differentiation will evolve as material circumstances evolve; and the general absence of bifurcation in the operational paragraphs of the agreement. Finally, the agreement adopted an innovative, hybrid legal form, with nonbinding emission targets, but a binding overall structure and binding rules. Barely noticed by most countries, but the wise handiwork of the small islands, another provision found its way into the Paris outcome, ¶21 of the decision, which invited the IPCC to prepare a special report in 2018 on the impacts of global warming of 1.5°C above preindustrial levels—a provision that became, three years later, the mouse that roared.

The new agreement was almost instantly hailed as historic, a landmark that sent a signal around the world to citizens, civil society, boardrooms, and governments that the leaders of the world had finally recognized climate change as a serious threat and come together to address it.

Jonathan Chait, writing in the *Intelligencer* two days after the deal was done, described it as a "triumph of international diplomacy shared by diplomats across the planet," and called it "probably the [Obama] administration's most important accomplishment," taking into account the monumental threat to the world that climate change poses. He noted the breakthrough between the West and the developing world represented by the 2014 US-China Joint Announcement. And he recognized the singular difficulty that the United States faced on account of the Constitution's Treaty Clause, requiring an inconceivable two-thirds of the Senate to approve a fully binding agreement, and how Paris worked around this obstacle. Acknowledging that "the hard part" still lay ahead, and noting that historic victories—citing the Emancipation Proclamation, the Thirteenth Amendment, and the World War II Lend-Lease Act—"are hardly ever immediate or complete," he argued that "the climate agreement in Paris should take its place as one of the great triumphs in history."[41]

The morning-after stories came a morning or two later, and in two varieties. Some correctly pointed out, as the *New York Times* did, that "now comes the hard part." The agreement did not, because it could not, mandate action by countries, so whether it would lead to the measures needed and commitment required would depend on political will around the world.[42] Others dismissed the agreement outright. as inadequate or worse because the NDCs were too weak or not legally binding.[43] But those criticisms failed to look at the real world of negotiations. Allowing countries to choose their own, nonlegally binding targets was the only way to get a deal done, and Paris complemented that reality with the ambition engine called out by Tony de Brum. Politics, whether domestic or the international kind called diplomacy, really is the art of the possible, and that understanding is built into every page of the Paris Agreement.

What is true about the Paris Agreement is that it gives no assurances. It sets goals for limiting temperature and emissions. It lays out a procedural path, with strong, binding rules meant to push countries in the right direction. And it makes a bet that the rising force of norms and expectations, together with the ambition mechanism of Paris itself, will make climate action important—for economic development and growth, national security, global standing, and reputation—prodding countries as well as companies, innovators, investors, and the consuming public to do better, do more, and hopefully do enough. That President Obama repeatedly referred to Paris as giving us a chance was no accident. That's what it does. But that is no small thing—giving us a chance, giving us hope, and setting in motion a virtuous chain reaction with the potential to create a decarbonized global economy and preserve a livable world.

In the end, why did the Paris Agreement come together? Part of it was that countries were ready for a deal, believed a deal was going to get done, were importantly influenced by the US-China Joint Announcement and feared what the specter of another failure might mean for the UNFCCC and multilateralism generally—no small matter. For the large majority of countries in the world that are not majors, the multilateral process is what gets them in the room, gives them a voice, and allows them, when those smaller voices join together, to have clout, as we saw in Paris. Moreover, the long lead time for the negotiation helped. Countries were familiar with the key issues. They had accepted the notion of bottom-up NDCs with top-down rules, accepted a revised formulation of the famous phrase "common but differentiated responsibilities," and knew this would be a legal agreement applicable to all, though almost surely not legal in all respects. If, in the autumn of 2009, the notion of a nonlegally binding accord with greater balance between developed and developing countries had caught parties by surprise in the weeks leading up to the Copenhagen conference, and in that way set the conference up for a fall, the situation in Paris was completely different. In addition, as my colleague Brian Deese wrote in *Foreign Affairs* in early 2017, the traditional political and economic headwinds to global action on climate change had been significantly eased by dramatic technological progress in developing cheap and effective clean energy.[44]

Beyond these background factors, many players made important contributions to getting the Paris deal done, and a few stand out. The European Union played a critical role. It had always been the conscience of high ambition, but throughout 2015 and in Paris, led by its tough, wily leader Miguel Arias Cañete as well as its lead negotiator, the United Kingdom's Pete Betts, it insisted on science-based goals, strong rules and procedures, and the end of bifurcation, and stood solidly with the United States and our key Umbrella Group partners on all of that. The fears that Pete and I talked about over a London dinner in February at the Cinnamon Club—that the United States would settle for weak rules in order to avoid the firewall (the EU worry) or that the European Union would go wobbly on the firewall in order to secure strong rules (the US worry)—never materialized. We stayed close together and formed a powerful alliance. And the European Union was also instrumental in reviving the progressive alliance that turned into the HAC. In the United States, Barack Obama's engagement and commitment, especially in the last three years leading to Paris, paid huge dividends, as did John Kerry's high-level, high-octane role throughout his tenure, all the way to the last week of the Paris conference.

And the United States and China came together as the pivotal bilateral partnership for getting the deal done. My friend Xie Zhenhua and I spent more hours talking to each other over seven years than I could ever count. And when Presidents Obama and Xi walked down the aisle together in Beijing in November 2014, they tied the two sides' fortunes together and made success in Paris an imperative. The senior US team met with China almost every day of the COP. We wrangled over ideas and language, but with a mutual understanding that we would find a way forward in the end. And China, with its influence in crucial quarters of the G77, had a unique capacity to bring along the most reluctant players.

The HAC, in which dozens of progressive and vulnerable countries, led by the strategic and inspirational Tony de Brum, made common cause with the European Union, United Kingdom, United States, Australia, among others, created a uniquely potent political force—more than a hundred countries of all sizes and shapes demanding not just a deal but a deal grounded in science and ambition. Pete's terse, strategic watchword from September—"if it's us plus vulnerables and progressives, we win"—was right on the money.

The HAC gave France the backing it needed to deliver an accord better than most expected.

Forces outside governments, including scientists, economists, businesses, and NGOs also played a valuable role in the years leading to Paris, and thanks especially to Laurence Tubiana, were given a more prominent place at the Paris COP than ever before. Scientists spoke most vividly through the IPCC's Fifth Assessment Report, released in 2013–2014. Businesses like Unilever, led by the indefatigable Paul Polman, and IKEA along with the new global network We Mean Business, founded in 2014, started advocating vocally for climate action, signaling to political and business leaders as well as their own employees and customers that there was powerful, mainstream support for climate action. And of course, activists and NGOs kept the pressure on for an agreement with ambition, accountability, and support for vulnerable countries, including by proposing and leading the People's Climate March in New York in 2014, some four hundred thousand strong, getting front-page coverage and sending a message to world leaders gathered a few blocks away in the United Nations.[45]

Finally, France itself did a masterful job all year long and in orchestrating the COP itself, delivering the essential substance that gave the agreement its historic lift, displaying skill, poise, finesse, and stamina in managing an always fractious assembly. The French understood the need to keep the trust of the parties, and ran a process centered on inclusion and the right kind of transparency. They chose a good group of ministerial facilitators to lead efforts on each major issue. Fabius and his team made it clear that they would listen to everyone, and that their door was open for one-on-one consultations. They subjected new versions of the text to extended comment in the Indaba process, where Fabius listened hard and jotted down notes about what different countries and groups needed. Crucially, Fabius along with Laurence and her team were able to distinguish real redlines from lines that, if necessary, could be crossed or at least blurred. And while always working from the negotiators' text, never having a "French text" in their back pocket, they in fact had full control over the evolving document, deciding what options stayed, what were removed, and what language needed to be modified and how. Finally, I came to understand by the time the final text emerged that France was guided by a kind of "true north." The French were pragmatic, as you have to be to find a viable path through the thicket of conflicting desires and demands in climate negotiations, but they were in no sense aimless. They knew where they wanted to go on the big things, and knew that they wanted to deliver an agreement with long-range ambition that aimed at net-zero emissions, included midcentury strategies to decarbonize the global economy, provided the prospect of more climate finance going forward, moved past the firewall for a regime pointed at the 2020s and beyond, and opened the COP process to the nongovernmental world as never before.

* * *

When I eventually left the plenary hall, I first kept a promise to Singapore's Vivian Balakrishnan by walking with him to the Chinese suite for a three-way photo of him, me, and Minister Xie, the two Goliaths that little Singapore did a better job of understanding and balancing than most anyone in the world. After that, Brian and I briefed the press, and then went back with our team to the airport Marriott to bask in the glow of the deal with some late-night bar food. Well past 1:00 a.m., as I heard the chatter of impending movement from our young team, I started to go up to my room, begging off the NGO party in Paris a half hour away, at which point Paul said, "OK fine, I guess you'll go the next time we do the biggest climate agreement ever reached." I paused for moment and said, "OK, that was a really good line, I'll be ready in five minutes." And off we went until the wee hours, rock stars at Le Players Club, which was packed wall-to-wall with climate activists in a whirl of enthusiasm, celebration, and selfies. On the way there, I phoned my sixteen-year-old son in Washington to brag about the unheard-of reality that his dad was going out clubbing at 2:00 a.m. A fun night after a day to remember.

10 The Road Ahead

The Paris Agreement entered into force in record time on November 4, 2016, establishing an essential foundation for global climate action. Without such an agreement, it would be hard to imagine how nearly two hundred countries could grapple effectively with climate change. But four days later, on November 8, Donald Trump was elected president of the United States, and four months after he took office, he announced, with much fanfare in the White House Rose Garden, that he was pulling the United States out of the Paris Agreement.

This was a blow to the new accord, but not a fatal one. Other countries were angry and frustrated, especially since the United States had shaped the Paris Agreement in critical ways that many countries accepted only because they knew it was vital for the United States to be on board. Yet the rest of the world carried on. Indeed, within two years, at COP 18 in Katowice, Poland, the "Paris rulebook" of detailed guidelines and regulations was virtually completed.

The climate change landscape looking ahead will be shaped by powerful forces. The punishing impacts of climate change will threaten our ability to safeguard our world. But extraordinary technological progress is providing us with the tools required to execute the rapid global transition we need in energy and land use. And yet there are substantial obstacles in our path, especially the difficult political economy of that transition, particularly the power of the fossil fuel industry. Whether we can overcome those obstacles will be one of the defining questions of the twenty-first century since the difference between executing the transition at roughly the speed and scale needed and failing to do so will have profound social, economic, political, national security, and even health consequences. How that question gets answered will depend on many factors, including a well-functioning Paris regime that helps drive ambition, the choices made by national governments to embrace strong climate action or not, the political lobbying of the fossil fuel industry but also, increasingly, of clean energy interests, and the views—the hearts and minds—of the global public.

The Paris Regime

In March 2019, at the first of a series of small annual meetings I attended of the "Friends of the Paris Agreement," hosted by Laurence Tubiana and Xie Zhenhua, Minister Xie said that "the Paris Agreement is our baby, and all of us must make sure that it will grow up in good health." He was right about that, and the record has been quite good in many respects, but there is room for significant improvement. As things stand, the agreement has passed through its toddler phase and reached the stage of promising adolescence, but it has a long way to go before it reaches maturity.

From the beginning, the Paris Agreement sent a message with real impact to business, markets, investors, and innovators that we were heading for a global low-carbon future. A December 2020 report by the London analytics firm Systemiq said, "The dynamics set in train since the Paris Agreement have created the conditions for dramatic progress in low-carbon solutions and markets over the last five years. The agreement—with its in-built 'ratchet' mechanism—laid out a clear pathway for 195 countries to steadily cut their reliance on fossil fuels. This shared direction of travel increased the confidence of leaders to provide consistent policy signals . . . creat[ing] the conditions for companies to invest and innovate, and for the markets for zero-carbon solutions to start scaling."[1]

The Paris regime lies at the center of the global response to climate change. (By "Paris regime," I mean the agreement itself, the processes and timetables it puts in place, and the actions countries take to implement it.) It mandates the "ambition mechanism" that the Marshall Islands' Tony de Brum called "the beating heart" of the agreement, including the two five-year cycles—one for countries to submit new ramped-up emission targets (NDCs), the other for a global stocktake to assess how the world in aggregate is doing in light of its collective goals and the latest climate science. It succeeded in delivering the Paris rulebook at COP 24 in Katowice and a reasonably effective first global stocktake at COP 28 in Dubai. It monitors the transparency regime as well as the committee on implementation and compliance. It calls on the world's leaders to gather together annually to focus on climate change, which in turn brings climate to the attention of their media and citizens back home. And largely as a result of Laurence Tubiana's vision, the Paris COP elevated the "nonstate actor" world of business, subnational governments, and civil society in the belief that their activities could help reduce emissions either directly or by catalyzing the efforts of others. Their participation in the annual COPs has grown rapidly, with the number of all registered attendees—negotiators, press, NGOs, international organizations, and business—swelling from some forty-five thousand in Paris in 2015 to nearly a hundred thousand at COP 28 in Dubai in 2023.[2] And the Paris regime has become the great convener, with the annual COP serving as the action-forcing event for delivering initiatives of all sizes and shapes, developed by parties and nonstate actors alike.

Science and Impacts

In the years following the Paris Agreement, the broadly accepted temperature limit shifted from well below 2°C to 1.5°C. The initial driver of this shift was the IPCC's special report, *Global Warming of 1.5°C*, published in 2018. Commissioned by the Paris Agreement and reinforced in 2021–2022 by the IPCC's *Sixth Assessment Report*, the 1.5°C report found that the sensitivity of the climate system to higher temperatures was greater than previously thought. This meant that severe impacts on human life and global ecosystems would occur at lower average temperatures than had been expected. The report's sobering analysis of the scientific evidence led previously reluctant countries to accept that 2°C could no longer stand as an acceptable temperature limit. This effective shift meant that the timeline for reaching net-zero greenhouse gas emissions was rolled forward from around 2070 to around 2050—a much tougher task.

This analysis from science, calling for faster, more intensive action, was amplified by the resounding drumbeat of extreme weather events around the world. The year 2023 left records for the hottest year ever in the dust, with the year's average global temperature some 1.35 to 1.48°C higher than the second half of the nineteenth century, depending on which expert body you were listening to—Copernicus, NASA, or NOAA.[3] This actually did not mean that scientists thought we had already all but reached the 1.5-degree goal, since what science considers to be a legitimate average global temperature usually depends on years of results, not just one. But still, these numbers were stunning. And then there were the jaw-dropping events. Phoenix, Arizona, had thirty-one consecutive days of 110°F or higher.[4] On July 12, *water* temperatures off the Florida Keys were over 90°F.[5] Canadian wildfires burned forty-five million acres, crushing the previous Canadian record of eighteen million, and sent smoke and haze billowing down to the American Midwest, with Chicago and Detroit posting the worst air quality numbers in the world on back-to-back days in June.[6] In August 2023, Brazil's winter, the temperature rose to 104°F.[7] In September, a catastrophic flood in Libya killed some four thousand people and displaced tens of thousands.[8] And so on. In 2022, China was scorched by a searing heat wave that lasted more than seventy days, affecting more than nine hundred million people from the southwest to the coastal east with sustained heat over 104°F.[9] That same year, the thermometer hit 116°F in Portugal,[10] while more than sixty-one thousand people throughout Europe died from heat-related stress.[11] In summer 2022, Pakistan recorded heat of 120°F[12] and was devastated by flooding that affected more than 33 million people.[13] Drought in the Horn of Africa from 2020 to 2023 was the worst in seventy years, threatening millions.[14] A World Bank report in October 2021 found that compared to the decade of the 1970s, the frequency of droughts in sub-Saharan Africa nearly tripled in the 2010s, the frequency of intense storms quadrupled, and the frequency of floods increased by ten times.[15] And rivers were running low all over the world, including the Colorado River in the western United States;[16] the Rhine, Loire, Po, and Danube in Europe; and Yangtze in China.[17] And climate change was the main culprit.[18]

All of these impacts, serious as they are, involve "linear" climate responses to the temperature getting warmer and warmer. But there is another kind of climate risk, referred to as "tipping points," in which changes that have been gradual suddenly cross a threshold and produce dramatic change. Scientists deploy a variety of metaphors to illustrate how tipping points work, such as the game of Jenga, which starts with a tower of blocks that players remove one by one, trying to keep the tower standing, until one block too many is removed, and the tower crashes down.

Tipping points that scientists have identified include a potential collapse of the ocean currents (known as the Atlantic Meridional Overturning Circulation) that cycle warm water up from the tropics to the northern Atlantic Ocean and back down again, keeping the climate in Europe temperate rather than very cold. Another is the potential conversion of the immense Amazon rainforest into a savanna, leading to a huge increase in carbon emissions from burning or cutting trees, catastrophic species loss, and dramatically drier conditions for South America as the "flying river" air currents that bring rain from the Amazon to much of the continent disappear.

Technology

Technological progress in recent years has been remarkable. Systemiq noted that the International Energy Agency (IEA) forecast in 2014 that solar prices would fall to five cents per kilowatt hour by 2050, thirty-six years later. In fact, it took six.[19] In the decade from 2012 to 2022, the cost of solar energy fell by 80 percent; the cost of onshore wind fell by 57 percent and offshore wind by 73 percent; and the cost of batteries, used to power electric vehicles as well as provide storage capacity for variable wind and solar energy, fell by 80 percent.[20] These falling prices led to a massive increase in the penetration of clean energy, which in a virtuous circle, further drove down prices. Global EV sales rose from less than 1 percent of cars sold in 2016 to 14 percent in 2022.[21] When the purchase price of EVs falls below equivalent internal combustion cars, projected to happen in the mid- to late 2020s in Europe, China, the United States, and India, vehicle markets are expected to change still more dramatically, with RMI projecting that EVs will account for two-thirds or more of global car sales by around 2030.[22]

Conventional modelers frequently failed to anticipate this dynamic green technology story. According to the Oxford Institute for New Economic Thinking (INET) in a 2022 paper, "Models . . . such as those used to inform the UN Intergovernmental Panel on Climate Change ('IPCC') and policymakers worldwide, have consistently and significantly overestimated the future costs of key clean energy technologies: "For example, the average value of . . . projected solar PV cost reductions for 2010–2020 in rapid decarbonization scenarios was 2.6% per year; the actual drop was almost 6-times faster at 15% per year."[23] What conventional modelers missed, and analysts like INET and RMI have understood, is the difference between linear thinking and so-called S-shaped curves, powered especially by learning. RMI's Kingsmill Bond said that as technologies grow, we see "falling costs, rising capabilities [learning by doing], rising consumer awareness, rising lobbying power, and more complementary technologies which all in turn spur greater scale." He noted that solar, wind, batteries, heat pumps, and green hydrogen all fit S-shaped curves.[24] And this speed is crucial because our chance to achieve our 2050 emissions and temperature goals turns on the pace at which clean energy supplants fossil fuels.

The energy storage needed to keep power flowing on grids that rely on intermittent wind and solar is also surging, with 2023 setting new records in the addition of grid storage.[25] In 2022, electric heat pumps, which supply both heating and air-conditioning, outsold traditional, fossil-fueled gas boilers in the United States, spurred by a $2,000 federal tax credit, while their sales increased by 10 percent worldwide and 40 percent in Europe.[26] Taken together, solar, wind, batteries, and heat pumps are the core enabling technologies to decarbonize electricity, transport, and buildings, which together account for over 50 percent of global greenhouse gas emissions.[27]

There is also intensive work going on around the world to accelerate the development of technologies that are not yet commercially viable, often for the so-called "hard to abate" sectors such as heavy industry, aviation and shipping. After the Paris COP, Bill Gates began to put serious money into this effort. By 2023, his venture capital firm, Breakthrough Energy Ventures, had invested in over eighty start-up companies developing a wide range of new technologies, from new processes to produce zero carbon steel and cement, to long-duration battery storage, alternative proteins, sustainable aviation fuel, and green hydrogen, among others. And that's just one among many venture capital and accelerator firms betting on the clean energy revolution. Work is going on in government labs as well, including on technologies, which, if they prove feasible, could be global game changers, such as nuclear fusion and geological hydrogen.[28]

Energy efficiency—the amount of work done per unit of energy—is also vital to delivering the emission cuts we need. It can be produced in many ways—improved standards for lighting, more efficient appliances and industrial motors, and better building codes and vehicle standards, among others. Electrification itself, such as electric rather than conventional vehicles or heat pumps rather than gas boilers, saves a substantial amount of energy too.[29] Efficiency is sometimes called the "first fuel" in the clean energy transition because of the role it plays in reducing emissions and cost along with enhancing energy security and because the energy you never need to use on account of greater efficiency costs nothing, emits no greenhouse gases, and creates no dependency on imported supplies.[30]

In its *Energy Efficiency 2022* report, the IEA noted "dramatically escalat[ing] concerns" over energy security and cost, triggered by Russia's February 2022 invasion of Ukraine. The IEA said that "while there are many ways for countries to address the current crisis, focusing on energy efficiency action is the unambiguous first and best response to simultaneously meet affordability, supply security and climate goals."[31] In its *Energy Efficiency 2023* report, the IEA underscored "a strong global focus among policy makers" on improving energy efficiency for precisely those reasons.[32] At the IEA's eighth annual Global Conference on Energy Efficiency in June 2023, forty-six countries, including the United States and those in the European Union, adopted the Versailles statement calling for a doubling of the annual rate of efficiency improvement, intended to achieve a 4 percent annual improvement through 2030.[33] A version of this goal was adopted at the Dubai COP in 2023.

Policy

Well-crafted government policy acts as a crucial enabler of technology development. Such policies include funding for research, development, and demonstration; feed-in tariffs of the kind Germany used successfully from 2000 to 2021 to assure producers of renewal energy that they had a buyer for any energy they produced; tax credits for the production of or investment in green technologies; consumer credits for buying products such as EVs or heat pumps to make them competitive in the marketplace; government procurement programs; and carbon pricing by way of carbon taxes or cap-and-trade programs.[34]

The big four global emitters have all deployed policies like these to drive the clean energy transition forward. In the United States, the Inflation Reduction Act (IRA), signed into law in 2022, is the heart of an aggressive climate program put forward by the Biden administration. Using a mix of grants, loans, and tax credits, the IRA provides some $370 billion in public money designed to leverage hundreds of billions more in investment—"government enabled but private sector led," in the words of my old colleague John Podesta, whom President Biden brought back to the White House in 2022 to quarterback IRA implementation.[35] Among a broad array of actions taken under the IRA's authority since it was enacted in August 2022 was a major announcement in March 2024 of $6 billion in grants, expected to leverage an additional $14 billion from the private sector, for industrial decarbonization. Industry is the most difficult sector to decarbonize, accounting for nearly a third of US emissions. Rebecca Dell, a leading expert on reducing industrial emissions, described the new program as "the most important thing that has happened in the history of industrial decarbonization."[36] The IRA, Biden's infrastructure and CHIPS Acts and proposed regulations to further limit emissions from power plants and light-duty vehicles can enable the United States to meet its climate target of reducing emissions 50 to 52 percent below 2005 levels by 2030.

The European Union has been at the forefront of climate policy for decades with strong international targets and domestic policy measures to help meet them. China is by far the world's largest producer and consumer of wind and solar energy, with more than one-third of the global total of both, and has built that position through extensive domestic policy measures.[37] And India has a suite of policies to follow through on Prime Minister Modi's pledges to rapidly expand India's renewable capacity. Early in his administration, he promised to build 100 gigawatts of solar capacity by 2022, more than thirty times what India had then, and 60 gigawatts of wind. At the Glasgow COP in 2021, Modi tripled down on his pledge, declaring that India would obtain half its energy from renewables by 2030, reach 500 gigawatts of solar and wind capacity, and reach net-zero emissions by 2070.[38] Collectively China (29.2%), the United States (11.2%), India (7.3%), and the European Union (6.7%) accounted for over 54 percent of global greenhouse gas emissions in 2022.[39]

Obstacles

Of course, there are real technological, economic, and political hurdles to reconfiguring the global energy and land use systems within a few short decades. We will need to make substantial investment in the grid; build out global capacity to procure critical metals and materials as well as manage the complicated geopolitics of supply chains for such elements; and accelerate the development of low-cost, low-carbon technologies in the hard-to-abate sectors, such as heavy industry, shipping, and aviation. But the most difficult obstacle is the global power of the fossil fuel industry, which still produces 80 percent of primary energy worldwide and has formidable political clout. Fossil fuel companies showed up in force at the Dubai COP ready to make useful peace offerings, but hell-bent on protecting their core business. Thus fifty producers, led by Exxon and Saudi Arabia's Aramco, promised to reduce methane emissions from oil and gas production to near zero by 2030—important if carried out because methane is responsible for some 30 percent of the current increase in global temperatures[40]—and cut emissions from their own operations to net zero by 2050.[41] But they were determined to avoid any reference to limiting the use of oil and gas, which accounts for the vast majority of the greenhouse gases they generate.[42]

The domestic political power of the fossil fuel industry also makes policy change difficult, even for something as obvious as reining in massive expenditures for fossil fuel subsidies, the last thing we need. In India, the central government owns around three-quarters of Coal India, the dominant producer, and derives considerable revenue in the form of dividends. The four poor Indian states that produce most of India's coal are dependent on the royalties they receive from coal production. And India's railways subsidize passenger fares with the earnings they get from hauling coal. In China, President Xi Jinping said in April 2021 that China would "strictly control coal-fired power generation projects, and strictly limit the increase in coal consumption" in the five-year plan running from 2021 to 2025, but he shifted course in 2022.[43] Following widespread electricity outages, caused in part by the reduction in hydropower that resulted from the 2022 drought, China granted permits to build over a hundred large-scale coal plants and started construction on six times more coal capacity than the rest of the world combined.[44] At China's twentieth party congress, in October 2022, Xi declared that energy security was the new focus and coal was a linchpin of that security.[45] That the world of energy had been disrupted by Russia's invasion of Ukraine doubtless reinforced Xi's pivot.

Even though the Biden administration is going full throttle to meet its emission reduction target, the United States is still the largest producer of oil and natural gas in the world, and with that economic reality comes political influence in Congress and relevant states. There is also a more local kind of drag in the United States, where multiple approvals are often required from states, cities, and counties for the new transmission lines needed to move wind and solar power to population centers, and a related problem, known as interconnection, that forces new wind and solar capacity to wait in a queue, sometimes for years, before getting added to their local grid. The Biden administration has worked intensively to clear up these problems, but the partisan divide in the United States makes this much harder than it should be.

The State of Emissions

The state of global emissions near the end of 2023 was not encouraging, though there were some hopeful signs. Based on figures from the 2023 Global Carbon Project's Carbon Budget report, overall CO_2 emissions from fossil fuels were still rising, with 2023 emissions around 1.1 percent higher than in 2022. Emissions from fossil fuels were falling in the European Union, United States, and twenty-six countries overall, representing 28 percent of global emissions, while they were rising more slowly in other countries. In China, for instance, responsible for nearly a third of global fossil fuel emissions, such emissions grew 8.4 percent per year in the decade 2003–2012, but 1.6 percent per year in the decade 2013–2022.[46] The growth of fossil CO_2 emissions in the aggregate of most other developing countries fell from 3.3 percent per year in 2003–2012 to 1.5 percent per year in 2012–2022. And the overall growth of CO_2 emissions from both fossil fuel and land use (agriculture and forestry) slowed to 1.4 percent per year in the decade 2013–2022 from 2.1 percent per year in the previous decade.[47] The Global Carbon Project also tracks possible carbon budgets for the world going forward, measuring the estimated gigatons of CO_2 that can be released before crossing a given temperature threshold. The 2023 report said the remaining carbon budget starting from January 2024 is 275 gigatons of CO_2 for a 50 percent chance of holding to a 1.5°C temperature increase, 625 gigatons of CO_2 to hold the temperature increase to 1.7°C, and 1,150 gigatons of CO_2 to hold the temperature increase to 2°C. Assuming emissions stayed at 2023 levels going forward, we would hit 275 gigatons of CO_2 in seven years, 625 gigatons of CO_2 in fifteen years, and 1,150 gigatons of CO_2 in twenty-eight years.[48]

Dubai and the Global Stocktake

The Dubai COP in 2023 was particularly important as the year of the first global stocktake, the every-five-year comprehensive assessment of collective progress toward achieving the long-term goals of the Paris Agreement. While the global stocktake (GST) dealt with many issues, its highest-profile focus was a battle between high-ambition countries and fossil fuel producers over whether the negotiated text of the stocktake should call for a phaseout or at least significant reduction of fossil fuels by midcentury (the high-ambition position), or should talk only about reducing emissions without mentioning the words "fossil fuels" (the pro-fossil-fuel position). Oddly, those words had never appeared in a COP decision going back to the Framework Convention. That the COP was hosted by the United Arab Emirates, a major oil and gas producer, and that the president of the COP, Dr. Sultan Al Jaber, also ran the Emirates' national oil company, ADNOC, complicated the equation.

The parties wrangled over the fossil fuel issue until past the last scheduled day of the COP, with Saudi Arabia, Russia, and their allies adamantly opposing any reference to fossil fuels in the final text of the global stocktake. In the end, the text called for "transitioning away from fossil fuels in energy systems, in a just, orderly and equitable manner, accelerating action in this critical decade, so as to achieve net zero by 2050 in keeping with the science." This language did several important things. It sent a clear signal about what the direction of travel should be for the production and consumption of fossil fuels. It sharpened the world's net-zero timeline to 2050 as compared to the less precise "around midcentury" contained in the 2021 Glasgow Climate Pact (¶22) and still-softer emissions goal in the Paris Agreement itself (Art. 4.1)—the equivalent of net zero "in the second half of this century."[49] And the GST text also called for both tripling global renewable energy capacity by 2030 and doubling the global average annual rate of energy efficiency improvements by 2030. All of these elements made the Dubai outcome significantly more ambitious than anything that had come before. In addition, regarding the next round of country NDCs, due in 2025, the GST text includes a strong paragraph (¶39) encouraging all parties to submit "economy-wide emission reduction targets," "covering all greenhouse gases, sectors and categories," "aligned with limiting global warming to 1.5°C, as informed by the latest science." The paragraph ends with the qualifier "in the light of different national circumstances," but was still the strongest ambition language ever applied to all parties.

The Saudis were visibly unhappy over the "fossil fuels" language, staying pinned to their chairs during the standing ovation when the text was adopted in the final plenary session of the COP. Still, fossil fuel producers secured a recognition that "transitional fuels," code for natural gas, "can play a role" in the energy transition, even though natural gas is a fossil fuel that in many countries, is barely less polluting than coal, owing to high rates of leakage.[50] And they came away with language they liked on carbon capture, utilization, and storage technology, which some see as a potential license to keep oil and gas exploration and production running on high, an unnecessary and counter-productive approach. As the IEA said in a 2023 report: "As clean energy expands and fossil fuel demand declines in the NZE (net zero) Scenario, there is no need for investment in new coal, oil and natural gas."[51] And an Oxford University report issued during the Durban COP found that heavy dependence on CCS to reach net zero emissions by 2050—as distinguished from a modest use to eliminate a relatively small amount of fossil fuel emissions that can't be avoided—would be "highly economically damaging," costing at least $30 trillion more by 2050, owing to the high cost of CCS technology.[52]

Overall, the reaction to the COP was mostly positive. The somewhat ambiguous language of the COP 28 text led some observers to claim that it had not effectively committed the world to phasing out of fossil fuels, but the UNFCCC's own press release had no doubt, saying that the agreement "signals the 'beginning of the end' of the fossil fuel era."[53] This did not mean that fossil fuel companies would suddenly shelve their plans for increased drilling. But it will make it increasingly difficult for such plans to be justified in the future. And this, in turn, will affect how businesses and governments behave starting in the near term.

Commentary from Nat Keohane, a veteran of the Obama White House and president of C2ES, the climate NGO, also gave high marks to private sector engagement: "What was striking was not only the quantity of participants: it was the quality of the discussions. Companies came not to show and tell, but to have granular, specific conversations about cutting emissions . . . deploying low-carbon technologies, scaling investment in climate action, and promoting adaptation and resilience to the impacts of climate change. . . . COP 28 was the conference when the focus turned squarely to solutions."

What Now?

During their 1985 Geneva Summit on the reduction of nuclear arsenals, US president Ronald Reagan and Soviet president Mikhail Gorbachev took a walk during a break in the negotiations. As Gorbachev recalled the story, Reagan abruptly said to him, "What would you do if the United States were suddenly attacked by someone from outer space? Would you help us?" Gorbachev said, "No doubt about it," and Reagan answered, "We too." There is a lesson here. The United States and Soviet Union were adversaries, armed to the teeth against each other. But as their two presidents imagined an attack from outside the boundaries of their shared planet, they agreed at once that they would help each other. The international community ought to look at climate change in roughly similar terms, as a threat that demands genuine partnership, something akin to a meteor headed toward the earth where we will have the best chance of pulling through if we all pull together. We are facing a genuine crisis. But not everyone seems to understand that.

In 2023, eight years after the Paris Agreement was concluded, too many countries still tried to pull backward toward the days of the firewall division between developed and developing countries as those countries were defined in 1992. China, for one, started referring to the "UNFCCC and its Paris Agreement." At the Friends of the Paris Agreement in Beijing in September 2023, the first draft of the chairs' summary, prepared by China, said, "It was stressed that the UNFCCC and its Paris Agreement are the core of international cooperation on climate change, with their principles, objectives, goals and provisions as the overall context and overarching framework." The point of this proposed approach was clear. The principles and provisions of the UNFCCC include the famous CBDR phrase, read by China and many others to mean the old firewall. The Chinese, together with their allies, wanted to restore as much of the old firewall spirit as possible to protect themselves against expectations that they should make deeper emission cuts than they preferred. But the Paris Agreement had fundamentally changed the nature of differentiation between developed and developing countries.

A focus based on proving how much a country should *not* have to do, especially a country whose annual emissions dwarf every other country including the United States, is not the right way to defend against a common threat to our planet—a very real one as distinguished from the fanciful threat that Presidents Reagan and Gorbachev imagined. The Paris Agreement ensures countries by way of its nationally determined structure that they cannot be forced to take action they do not want to take. But just as important, indeed essential to the agreement's normative commitment to ambition, it explicitly calls for a "do your best" approach to NDCs: each party's successive NDC "will represent a progression beyond the Party's current nationally determined contribution and reflect its highest possible ambition" (Art. 4.3). Article 4.3 ends with a reference to CBDR, but it is the modified version that adds "in the light of different national circumstances," a dynamic rather than static recognition of differences, since a country's "national circumstances" are continually evolving. In short, the Paris Agreement makes it clear that countries should never be expected to do more than their best, but should never be excused if they do less. If the Paris Agreement is to "grow up in good health," we need to take care to preserve this fundamental norm so that the Paris Agreement can take care of us.

High Ambition

To make the Paris regime as effective as it should be, there needs to be an ambition coalition of the kind that was so pivotal in Paris when Tony de Brum led it. The HAC still exists as a grouping of countries, but it lacks the significant impact it had in Paris where it included a wide range of countries, "big and small, rich and poor," in Tony's words, and used the broad-based power of its voice to insist that all countries pull their weight, defending the agreement's core commitment to ambition. If a coalition including members of the Climate Vulnerable Forum (fifty-eight countries), island states, Latin progressives, Europeans, United States, and most other developed countries locked arms, it would have real power in pressing for high ambition, just as it did in Paris. But to make this kind of coalition possible, and for good and important reasons of its own, solving the problem of financial assistance is essential.

Financial Reform

For a long time in climate negotiations there has been an angry, trust-depleting relationship between developing and developed countries over the question of finance. At the Copenhagen COP in 2009, developed countries made their famous pledge to mobilize $100 billion a year by 2020, which was initially well received, but then fed an increasing loss of trust after a multiyear failure by donors to meet the pledge. Climate-related anger has also been reinforced by the perceived failure of developed countries to treat the Global South fairly on issues including access to vaccines during the COVID pandemic and a growing debt crisis that forces many poor countries to spend more on servicing their debt than on basic needs.

Reports prepared in 2022 for the Bali G20 meeting and COP 27 in Sharm El-Sheikh, and in 2023 for the Indian G20 and COP 28 in Dubai, all focused on reforms to the World Bank and other multilateral development banks (MDBs) to generate much larger flows of capital to developing countries. These reforms involved more innovative ways to use their existing MDB balance sheets, a much greater use of their resources to derisk private sector investment, and an increase in paid-in capital from country shareholders —where every $1 of new capital for the World Bank could catalyze some $15 in lending.[54] Mia Motley, the prime minister of Barbados, announced her own "Bridgetown Initiative" in 2022, focusing not just on the MDBs, but also spreading natural disaster clauses in debt instruments as well as the role the IMF could play by increasing its rapid credit facilities and using so-called special drawing rights (the IMF's reserve currency) to expand the concessionality of, and access to, climate-related assistance.

All of these efforts recognized that very large sums of money were needed to propel the clean energy transition and build resilience in developing countries—something on the order of $1 trillion per year from international sources. The reports also recognized that recipient countries have their own important role to play by mobilizing substantial domestic resources, creating more appealing environments to attract foreign investors, and developing country-wide investment plans. Of course, developing countries differ widely from one another in terms of their needs and capacities. For some, attracting foreign investment will be key. Others, less developed, will need to rely more heavily on concessional financing.

This level of reform won't be easy, but the cause is critical. A great many developing countries need the assistance, the capacity for countries in the Paris regime to work constructively together depends on it, and addressing the finance problem effectively would have clear geopolitical benefits since it would strengthen relationships between the Global North and Global South. Larry Summers, former Treasury secretary and former chief economist of the World Bank, and noted Indian economist N. K. Singh, who together oversaw preparation of the "Triple Agenda" report for the 2023 Indian G20, wrote in an accompanying op-ed that "even as developing countries face much larger financing needs to meet development and climate goals, MDBs' disbursements have not kept pace, and the degree to which they now transfer resources to developing countries is unacceptably low."[55]

Meanwhile, in June 2023, a new president, Ajay Banga, took the reins of the World Bank. This was a welcome change, engineered by the Biden administration. Banga is an Indian-born US citizen with a strong financial background, credibility in the Global South, and the wherewithal to lead the profound cultural change that will be needed in the World Bank and other MDBs to take them in a new direction.

What we know is that the donors of the world have to deliver the goods if we are to strengthen the Paris regime's capacity to help drive global decarbonization and resilience at the same time that we repair geopolitical damage in an alienated Global South. And making important progress toward solving the finance problem holds out the best hope of evolving the Paris regime toward a productive partnership of countries pulling in the same direction rather than sniping at each other.

The MEF

The United States launched the MEF process in March 2009 in the belief that it would be useful for leading developed and developing countries to meet several times a year at a high level to talk frankly in a collegial setting. It was effective then and can be effective again. Of course, the context is different now. The MEF during the Obama years was overwhelmingly focused on the climate negotiations, which led, by fits and starts, from the Copenhagen Accord to the Paris Agreement. The MEF's core mission now should be to drive rapid decarbonization and build resilience against climate impacts, with leaders of MEF countries meeting every year in person to share, discuss, and debate views about what should and must be done. To prep for the leaders session, there should be at least two in-person sherpa meetings led by relevant ministers or senior advisers. The Biden administration has, in fact, hosted several leader-level MEF meetings starting in spring 2021, but these meetings have been remote, rather than in person, initially because of COVID, and have been focused more on announcing specific initiatives than on engaging in a concerted discussion about what we can and must do collectively to address the crisis. The special sauce of our "first-generation" MEF meetings was their intimate, back and forth, interactive character, and we should strive to re-create that, certainly in the sherpa meetings and as much as possible in the leader meetings.

I would also start every leaders meeting with a concise and compelling report, with good visuals, on what the latest science is telling us, delivered by one or two noted scientists of high credibility and with excellent skills in communicating to nonscientists. It is a mistake to assume that all the leaders of MEF countries understand the climate risks we're facing. They certainly know about climate change at a broad level. But whether they understand how serious the climate threat is, and how much and how quickly we need to decarbonize and build resilience—even recognizing the political economy difficulties of such action—is a very different question. Such a yearly review would set the stark parameters within which the rest of the discussion would take place. In addition, I would expand the MEF's membership by adding Turkey, Saudi Arabia, Argentina, and the African Union. It is time to have the membership of the MEF mirror the G20, and doing that would help in scheduling in-person leaders meetings as well. Finally, I would include on a rolling basis around five additional participants, among them small islands and vulnerable countries, whose perspective and reputation would make them useful participants. If, for any reason, including US politics, the United States were not in a position to lead, others could step up.

Of course, we have seen what a United States devoid of political leadership on climate change looks like—it happened from 2017 to 2020—and if that kind of thing were to happen again, the most important response from the Paris regime would be to once again carry on. In addition, the US Climate Alliance—a bipartisan group of state governors, currently twenty-five strong, created in the immediate aftermath of Trump's Paris withdrawal announcement—should appoint an envoy who can track what is going on year by year in the international sphere, including the annual COPs. The alliance would never be in a position to literally represent the United States, but should the United States again be led by a climate denier, the alliance could attend COPs and other relevant gatherings, talking knowledgeably about what US states, cities, businesses, and civil society are doing to combat climate change and engaging with other countries on how subnational governments in the United States might collaborate with subnational actors in their countries.

* * *

Beyond the MEF, groups of countries are collaborating to drive stronger climate action. At the 2022 G7 meeting that it hosted, for example, Germany proposed the creation of a voluntary climate "club" whose members would aim to advance the green transition in a variety of ways, including focusing on the decarbonization of different sectors.[56] And ad hoc groupings of countries, companies, international organizations, philanthropies, or NGOs have already come together to pursue initiatives and should enhance this activity. All such efforts should be welcomed and reported on at the annual COP.

The United States and China

The relationship between the United States and China in the years leading to the Paris Agreement was the most important bilateral climate relationship in the world. Each loomed large in both size and influence, with the United States as the world's largest cumulative emitter, and China as the largest emitter and the second largest cumulatively. Each occupied a great deal of space in the other's consciousness on a range of issues, including climate change. Each knew that finding a way forward with the other on disputed climate issues would usually lead to a successful result in the multilateral context. And many countries take a signal from the US-China relationship; if the United States and China are working constructively together, it boosts confidence throughout the broader Paris regime.

But in the years since the Obama administration, an already difficult and contentious relationship has become more so. A widely held view on both sides of the political aisle in the United States has been that China's objective is to unravel the post–World War II international order built on the rule of law, a market-based global economy, and the US alliance system. For his part, Chinese president Xi has said that US policy toward China amounts to "comprehensive containment, encirclement and suppression" of China.[57]

In 2020, as preparations began to be made for a potential Biden administration, both the United States and China started thinking about how to manage a resumed climate relationship. I took part in at least six high-level track two dialogues on climate change together with John Podesta and often former Australian prime minister Kevin Rudd, president of the Asia Society, while Minister Xie, then at Tsinghua University, led the Chinese side. Though we didn't then know how strained the US-China relationship would be in a Biden administration, it was clear enough that the relationship would be difficult. Since climate change was too important to ignore, we talked of how climate engagement between the two countries would have to proceed in its own "lane," if necessary.

Once the Biden administration began and the broader relationship between the two powers turned visibly raw, progress on climate change became difficult. The two sides produced two modest, but useful joint statements in 2021, but the challenges of trying to make progress in the midst of a snarling overall relationship were apparent. In September 2021, Foreign Minister Wang Yi told John Kerry, Biden's climate envoy, that climate cooperation could not be separated from the larger relationship: "The U.S. side wants the climate change cooperation to be an 'oasis' of China-U.S. relations. However, if the oasis is all surrounded by deserts, then sooner or later, the 'oasis' will be desertified."[58] Less than a year later, after Speaker of the House Nancy Pelosi's visit to Taiwan in August 2022 predictably infuriated the Chinese government, China suspended cooperation with the United States in several areas, including climate. Presidents Xi and Biden finally met in person for the first time during the November 2022 G20 gathering in Bali, twenty-two months after Biden took office, with the intention to start thawing out the relationship. But that tentative effort was knocked off course in late January 2023 when a Chinese spy balloon drifted over the United States and the US shot it down. The effort to improve relations resumed in the summer of 2023 when Secretaries Blinken, Yellen, and Raimondo along with Special Envoy Kerry all visited Beijing. And Presidents Biden and Xi had another useful meeting in November of that year.

Also in November, the two countries resumed climate cooperation when Secretary Kerry and Minister Xie, who had a very cordial, long-standing relationship, met for three days at Sunnylands in California. Their meeting produced the Sunnylands Statement, which included a joint commitment to reduce methane and jump-start the joint working group on climate action, first agreed to at the Glasgow COP in 2021, as well as common understandings about important elements of the global stocktake. Although China supported the Saudis and others in opposing a reference to fossil fuels at COP 28, the US-China climate relationship functioned reasonably well in Dubai, working toward common ground on textual issues and announcing at the end of the COP that both countries would again update their mid-century strategies, inviting other parties to do the same.

But the nature of the clean energy transition itself introduces additional complications. As the world moves rapidly toward reliance on clean energy—solar, wind, batteries, electric vehicles, electric heat pumps, and so on—the demand for critical minerals like lithium, cobalt, nickel, and many more will also rapidly rise, requiring a large-scale increase in supply through expanded mining, processing, and recycling.[59] China holds a dominant commercial position, controlling some 85 percent of global critical mineral processing and 60 percent of global production.[60] It also has an imposing capacity to produce green technology at an enormous, worldwide scale and the determination to use that capacity to help power their economy.

The desire of the United States, Europe, Japan, Australia, and others to diversify their clean energy supply chains makes sense as a matter of both building their own industrial capacity and avoiding dependence on any single country that could, at some point, weaponize its power over energy. Russia taught us that lesson when it cut off gas supplies to Europe in 2022, and China has its own legacy of deploying coercive trade measures—ask Australia, Norway, Japan, Korea, or even the National Basketball Association. But the challenge faced by those pursuing diversified supply chains is real—China produces well and cheaply.

Managing all of this while also seeking a constructive climate relationship with China will be tricky but important. In a world destined to become more and more endangered, the United States, Europe, and China, among others, need to maintain genuine climate collaboration.

The Sociopolitical Tipping Point

The central dilemma for climate change is one of political will and human motivation. Of course, the task itself is enormous—rapidly transforming the global economy from one run on fossil fuels to one run on clean energy, while also preserving natural capital and limiting emissions from land use—and it would be tremendously challenging even if all countries around the world were fully on board. Yet we have the capacity to do what is required. We have the innovative talent. We know what policies to deploy to hasten the transformation. And we can afford it, including the imperative of providing a "just transition" for those who need it, such as workers in a shrinking fossil fuels industry.

For those who think the transition we require is just too hard, recalling the early days of the COVID pandemic in 2020 is instructive. No one would have imagined, before the pandemic, that leaders would shut down entire cities of five to ten million people virtually overnight. How could that be done? What about the economy? What about the backlash? It would have seemed absurd. Until it didn't. All hard questions of this magnitude need to be considered by way of a "compared to what" analysis. Compared to the nightmare vision of a plague raging through their cities, political leaders stepped up and did the unthinkable. Of course, climate change does not present that kind of terror, with its lethal immediacy. And yet judged by the monumental dangers posed by climate change, a "compared to what" analysis should be deployed to assess how strong a set of steps leaders should take.

Political leaders inevitably worry about jobs, economic growth, national security, and the next election, and they pay attention to securing the support of powerful interests. Business worries about the bottom line, producing results for shareholders. And progress is slowed down by those who think of themselves as the "grown-ups" and believe decarbonization at the pace we need is unrealistic. Moreover, for ordinary people, it is natural not to want to face what nature is telling us; or not to grasp the urgency of the climate threat when most days don't seem so threatening; or not to be able to focus on this problem when so many immediate ones command their attention; or to feel helpless in the face of a problem that seems so daunting and a fossil fuel infrastructure that seems so vast; or to prefer just doing what their community has always done rather than confront change; or to assume that one way or another, we will muddle through. And of course, in some cases—consider Trump and Brazil's Jair Bolsonaro—political leaders are willfully ignorant and play to a populist crowd eager to scorn climate change and most anything else favored by the other side in polarized countries.

What we need to build toward, with all the commitment and speed we can muster, is a broad change in hearts and minds—normative change that can demonstrate to political leaders that their political future depends on taking strong, unequivocal action to protect our world. New norms may seem at first blush like weak reeds to carry into battle against powerful defenders of the status quo, but norms can move mountains. They are about a sense of what is right, what is acceptable, what is important, what we expect, and what we demand. This kind of normative shift in public attitudes has already started in regard to climate change and needs to gain momentum. And it has happened before. Broad public concern that the diminishing ozone layer in the atmosphere could give rise to skin cancer led to the negotiations that produced the successful Montreal Protocol in 1987. In 2010, after the US embassy in Beijing started to publish real-time, accurate information on dangerous fine particulate air pollution ("PM

2.5"), Beijing's citizens grew angry, began demonstrating, and caught the attention of China's leadership. Under public pressure, the Chinese government, which hates social unrest, started cleaning up Beijing's air. The original Earth Day, April 22, 1970, was the product of a gathering environmental consciousness spawned by Rachel Carson's famous 1962 book, *Silent Spring*, on the dangers of pesticides and by the infamous 1969 incident of Ohio's polluted Cuyahoga River catching on fire. Later that year, Senator Gaylord Nelson of Wisconsin called for a national teach-in on the environment, hoping to catch the activism young people had demonstrated in connection with the Vietnam War and the civil rights and women's movements. It worked. Some twenty million people nationwide attended thousands of events, and this demonstration of galvanizing public demand led in short order to the December 1970 creation of the Environmental Protection Agency, the Clean Air Act of 1970, Water Pollution Act of 1972, and Endangered Species Act of 1973, all during the Republican presidency of Richard Nixon.

What that original Earth Day represented was not just norm change but rather a tipping point in environmental concern—the kind of tipping point we need to reach on climate change as well, given the increasingly harsh response of an earth whose delicate balance has been ignored for too long. That tipping point will inevitably come, but once again, timing is crucial. We need it sooner, not later.

Many factors can combine to drive norm change and build toward a tipping point. The crescendo of extreme events around the world year after year plays an important part. People live through these events, see their family or friends live through them, watch them on television or smartphones. The backdrop of their world changes. The technology revolution embodied in wind and solar energy and EVs is moving fast, with rapidly declining prices and expanding reach, so that what were seen as niche products not long ago have become mainstream. The movements of markets are going to bolster renewable energy at the expense of fossil fuels. The IEA said in its 2023 *Net Zero Roadmap* that demand for fossil fuels will peak in this decade, and depending on policy, could even decline by as much as 25 percent.[61] And as the RMI fossil fuel and clean-tech analyst Kingsmill Bond has long argued, peaks bring disruption: when growth stops, incumbents face overcapacity, lower prices, and stranded assets, and when markets see decline set in, they don't wait long before disinvesting in the old and reinvesting in the new.[62] All such signals, picked up by investors and other market watchers, reinforce the sense that clean energy is attractive, that it works, that it's growing, that it's our future.

Norms are also built up by the actions of major companies across a broad range of sectors that are moving to embrace green solutions by acquiring their energy from renewable sources, setting voluntary targets to reduce their own emissions, selling green products like electric cars or plant-based burgers, and marketing their green bona fides to consumers. Business coalitions such as We Mean Business, the Energy Transition Commission, Ceres, and Business for Social Responsibility are helping to drive the green transition. More than eight thousand companies have pledged to cut their greenhouse gas emissions to net zero by 2050.[63] This is not quite as impressive as it sounds, since many have little idea how they are going to get there. Still, the rush to get on the right side of climate change at least demonstrates that companies see this as important to their customers, shareholders, employees, and the new young talent they are trying to recruit.

And of course, civil society has a crucial role to play. It has played an active and important role in calling for climate action for many years and will continue to do so. The original Earth Day in 1970 happened in a societal moment that isn't easily replicated, but it does teach that there is still more that could be done in filling the streets and campuses with young people who see the danger of climate change for what it is, see that they are the ones who will inherit a broken world if we fail to fix it, and have little patience for those who bury their heads in the sand.

There is also work to be done in talking to people from conservative communities traditionally viewed as hostile to climate action, particularly younger people who, as polling shows, accept the reality of climate change and the need to act on it far more than their elders.[64] In April 2023, I met an activist with a PhD in physics from a third-generation coal mining family in West Virginia whose assignment at the NGO he worked for was to advocate for a just transition for coal workers and communities, while also advocating with those workers and communities to embrace the clean energy transition. There should be a corps of such people in the United States gathered together in bipartisan effort.

When normative change goes far enough, it can reshape politics. Some political leaders, of course, are already fully committed to meeting the climate threat, and when they are, they can bring much of the public along with them. But most leaders are likely to go all-out on climate change only at such a time as they believe it will be bad for their political health if they do not. That means receiving stronger signals from the public that it demands much more aggressive climate action. It means hearing from companies willing to speak up loud and clear, not sotto voce, and lobby government officials to take strong climate action. As the clean energy side of business builds its own political clout, that will help as well. Smart legislation like the IRA is premised on catalyzing investment in green technologies, including in "red" states, and this in turn is creating a new cadre of companies and employees lobbying in support of the clean energy agenda. In a striking example of how this can work, a battle played out in the Texas legislature in June 2023 between the anticlimate governor, Greg Abbott—who proposed a set of bills intended to kneecap renewable energy in Texas—and a coalition of environmentalists, industry organizations, and business groups bent on protecting Texas's fast-growing renewable energy industry against Abbott's proposals. The good guys won.

Conclusion

The Paris Agreement, while far from perfect, was an enormous achievement and is now the principal international vehicle for meeting the climate challenge. The Paris regime must continue to play a key role in driving high ambition toward meeting our emission and temperature goals. Countries will continually need to be prodded to adopt ever-stronger emission targets, build resilience, and support those who need assistance. The Paris regime will also need to do its part in making climate action a norm throughout the world, helping overcome our reliance on fossil fuels, and sending a message to leaders that failing to act vigorously on climate change is an unacceptable strategy.

Climate change is as serious as scientists say and as nature is telling us. No one who has belittled the issue or assumed that holding the temperature increase to 3°C or 2.5 or 2.0 would be good enough has turned out to be right. The warnings from science have always been on target except when they have *underestimated* the impacts and danger. We should accept that 1.5°C is the right goal and should stay as close to it as humanly possible if we can't meet it all the way. Look at how extreme the climate impacts have been around the world in recent years with the temperature increase at "only" 1.1°C or 1.2°C.

We should never slip into the comfort of thinking that we will be able to just muddle through. That is often true, in different circumstances, but not here. The risks are too serious, and the size and scale of what could go wrong with our world are too dire. As Jared Diamond demonstrated in his book *Collapse*, humans have not always muddled through in the face of environmental risk. Civilizations have disappeared because they lacked the wherewithal to both recognize and address looming environmental crises.

And we should pursue research and development of strategies, such as the removal of CO_2 from the air, never to justify the continued production or consumption of fossil fuels, but because despite our best all-out efforts to accelerate global decarbonization, we are likely to still need those strategies.

But there is also a real basis for hope. The speed of technological progress is amazing, and even though we have not made enough progress yet, we can still reach our goals or come close to them. Consider the intriguing forecast in the IEA's *World Energy Outlook 2023* report concerning its Announced Pledges Scenario, which sits in the middle between its Stated Policies Scenario and Net Zero Scenario. The Stated Policies Scenario is based on what governments are actually doing; the Announced Pledges Scenario adds what governments have promised, including NDCs and longer-term pledges, such as to achieve net zero; and the Net Zero Scenario adds the policies that IEA's *Net Zero Roadmap* believes governments should put in place. Here's what's interesting: the IEA projects in its 2023 report that under the Announced Pledges Scenario, the global average temperature will rise 1.7°C by 2100, whereas just two years earlier, in 2021, it projected that global temperature would rise 2.1°C by 2100.[65] A move from 2.1°C to 1.7°C in two years is an improvement of nearly 20 percent, and a highly consequential 20 percent. It indicates the speed with which the IEA sees the clean energy transition moving. And we have the capacity to do much better than the Announced Pledges Scenario.

Large-scale sustainable development and resilience finance has to be a top priority looking ahead, and it must be approached with realism and expertise, taking advantage of the World Bank and other MDBs, the IMF, and national development banks. The way forward is not just to come up with big demands but rather to develop a system that actually produces capital where it's needed. This is the way we can help move developing countries forward and open the door to a more collaborative, ambitious partnership among the parties to the Paris regime.

The task of building broad, engaged, committed support for climate action is essential, establishing a powerful norm about the necessity of the net-zero transition, changing hearts and minds, and remembering the twenty million people in the streets on the original Earth Day and what they accomplished. At the end of the day, climate change is about the fate of our children, our families, our communities, our countries, our world. Governments, businesses, and civil societies can do what must be done. And when anyone says the goals are too hard, too difficult, would cost too much, would require too much effort or demand too much change, ask them: compared to what?

Acknowledgments

This book is my firsthand account of the seven-year road to the Paris Agreement on Climate Change, but the book could only have happened with the support, participation, advice, wisdom, and strategic counsel of scores of others.

Many people helped in a variety of ways with the book itself. I spent hours early in the process with Bo Lidegaard in Copenhagen, talking through events in the seminal year of 2009 as well as the broader structure of the book. Bo was the original champion of this project, read and gave me comments on the manuscript as it took shape, and continually pushed and prodded me in the best ways. Michael Jacobs, a wonderful writer, thinker and strategist, gave me invaluable advice and suggestions from broad structure to line edits as the manuscript moved along. I had many useful conversations with colleagues from around the world on different aspects of the story, including Luiz Figueiredo, Andre do Lago, Jairam Ramesh, Alf Wills, Emmanuel Guérin, Anne-Sophie Cerisola, Jimmy Fletcher, Farhana Yamin, Jake Werksman, and Artur-Runge Metzger, among others. And I talked regularly with Pete Betts, asking questions on matters large and small, and he was, as usual, lucid and on point with his answers as well as unfailingly generous with his time, even as, in his last eighteen months, he fought, with his usual tenacity, against a disease he knew could not be beaten. I am also grateful to the others who read all or parts of the manuscript and gave me valuable comments, including Jo Tyndall, Burhan Gafoor, Bill Antholis, the peer review readers at MIT, and Strobe Talbott, who years ago recruited me to the Brookings Institution as a place where I could work on climate change and write my book. Harold Koh, a former Dean and current professor at Yale Law School, as well as the Legal Advisor at the State Department during Hillary Clinton's tenure, was endlessly supportive and played a special role, recruiting me soon after the Paris deal was done to give a lecture and then teach a course with him and Doug Kysar at Yale on how the Paris Agreement came to be. The notes I prepared for those classes served as a kind of first outline for the book. Harold lined up excellent research assistants for me too, and I greatly appreciate the help provided by Natasha Brunstein, Justin Cole, and Griffin Black. I am also grateful to the Records-Archives team at the UNFCC for their invaluable help in accessing video records of COP meetings.

Two former science advisers from my team at State, David Reidmiller and Ben Zaitchik, together with Ben's assistant Talia Shadouri, helped me understand what I needed to know about new developments in climate science and climate impacts. I also benefited enormously from the talented clean energy team at RMI, where I served on the board for five years. Jules Kortenhorst and his successor as CEO, Jon Creyts, were full-fledged supporters of my project, and connected me to a number of their clean-tech specialists. Kingsmill Bond and his partner Sam Butler-Sloss, also from RMI, taught me about the power of learning by doing and S-shaped curves that tell the story of a quickening energy transition.

I had the privilege of working with a terrific team at the State Department. They managed to be smart, tough, and diligent while maintaining a wonderful esprit de corps that made me (almost) always happy to be spending hours in our windowless conference room planning strategy. It is in the nature of a book like this that most of them cannot be called out by name but that's not for lack of talent or importance. The are just too numerous to list. But they were outstanding. Indeed, I never went to an annual COP meeting without concluding that the US team was best in class. Sue Biniaz, who figures prominently in the book, was a brilliant strategic adviser, confidant, and my partner throughout, and I owe her a huge debt of gratitude. The unsinkable polymath Jonathan Pershing, my deputy in the first four years, was a priceless asset, managing the team on a day-to-day basis and helping to teach me what I didn't know. Trigg Talley, who ran the Office of Global Change at the State Department and became my deputy in President Obama's second term, was smart and measured, knew everyone in the negotiators' world, and had fine and nuanced strategic judgment. Christo Artusio was Trigg's deputy in the Office of Global Change and then took over from Trigg when Trigg became my deputy, and was an accomplished negotiator and manager. So was Kim Carnahan, who was an essential player during my tenure and then stayed on with Trigg to keep the office afloat during the tough years from 2017 to 2021. Trevor Houser, who handled the China portfolio for me in 2009, came back on a contract in 2014 to work on the special China project that year and continued over the years to help me think through a variety of climate issues. Dan Reifsnyder, a climate veteran, ran an adjacent environmental office at State, but was an active player on the climate side as well, a calm font of knowledge and wisdom, and a chair of the international negotiations in both 2011 and 2015. I was also served by talented chiefs of staff, including Pete Ogden, Jen Austin, Franz Hochstrasser, the endlessly upbeat Clare Sierawski, and the one and only Kareem Saleh, with whom I had some of my best adventures on the road. And Ben Kobren was an unbeatable press and communications adviser. I also always worked closely with the White House, and benefited greatly from the judgment and advice provided by Mike Froman, John Podesta, Brian Deese, and Caroline Atkinson. Reggie Love, President Obama's indispensable aide, helped piece together important moments of the President's dramatic and consequential day at the Copenhagen COP in 2009.

I worked with and learned from ministers and negotiators from all over the world, but some stand out for their contribution to the larger process as well as making me more astute and effective than I ever would have been without them. They include Penny Wong and Greg Combet (Australia); Selwin Hart (Barbados); Luiz Figueiredo, Andre do Lago and Izabella Teixeira (Brazil); Xie Zhenhua (China); Miguel-Arias Cañete, Jake Werksman, and Artur Runge-Metzer (European Union); Laurent Fabius, Laurence Tubiana, Emmanuel Guerin, Anne-Sophie Cerisola, and Jean-Paul Levitte (France); Karsten Sach (Germany); Jairam Ramesh and Dr. J. M. Mauskar (India); Franceso La Camera (Italy); Seyni Mafo (Mali); Tony de Brum (Marshall Islands); Luis Alfonso de Alba and Patricia Espinosa (Mexico); Tim Groser and Jo Tyndall (New Zealand); Manuel Pulgar-Vidal (Peru); Marcin Korolec (Poland); Ed Miliband, Ed Davey, Pete Betts, and Michael Jacobs (United Kingdom); Vivian Balikrishnan, Burhan Gafoor, and Kwok Fook Seng (Singapore); Alf Wills (South Africa); Teresa Ribera (Spain); Jimmy Fletcher (St. Lucia); and Christiana Figueres and Halldor Thorgeirsson (UNFCCC secretariat).

I have also benefitted from Dan Bodansky and Lavanya Rajamani, who for many years have thought hard and written astutely about the climate regime, and from Eliot Diringer at C2ES, who almost always had his finger on the pulse of where the negotiations had gone and where they were going.

There have been many people who helped me learn the ropes in Washington, and had confidence in me after I jumped from law to politics and policy, ultimately helping prepare me for the opportunity to lead US climate negotiations for more than seven years. John Podesta has been, more than anyone, my professional mentor in Washington. We have worked together often and he has over and over again opened doors for me. And others have played important roles, including Tom Donilon, Doug Sosnik, Jim Steinberg, Larry Summers and Neal Wolin. Kevin Rudd, whom I first met in Copenhagen when he was Australia's Prime Minister, has been a friend and supporter for years and my best teacher about China. I will always be grateful to Senator Patrick Leahy for giving me my first job in government. And to Bill Clinton's Chief of Staff, Erskine Bowles, who threw me into climate change in the first place in 1997. I learned a lot at my first COP that year, in Kyoto, from Stu Eizenstat, the head of the US delegation, and Vice President Al Gore, and learned over the years from Tim Wirth, the State Department's climate leader in the 1990s.

Of course, I also had the good fortune to work for two distinguished secretaries of state, Hillary Clinton and John Kerry, and to work with their top aides, Jake Sullivan on the Clinton side, and David Wade and Jon Finer on the Kerry side. And I had the privilege to serve President Barack Obama. The commitment of the President and Secretaries Clinton and Kerry is what made possible the seven-year journey culminating in the Paris Agreement.

When the time came to learn about navigating the process of getting a book published, I had excellent assistance. By way of my friend Lissa Muscatine, who referred me to Professor Richard Lazarus at Harvard, I found Thomas LeBien of Moon and Company, who gave me confidence in the project and taught me how to make my way to the next stage. I am grateful to Suzanne Maloney at Brookings for connecting me to my agent, Bridget Matzie, and to Bridget for giving me great advice on sharpening my book proposal and seeking a publisher, never failing to be upbeat and confident. And I am of course grateful to my excellent publisher, the MIT Press, and my editor, Beth Clevenger, who has been terrific and a pleasure to work with from day one.

Finally, I am eternally grateful to the bedrock support of my wife, Jen, and our three boys, Jacob, Zachary, and Ben. Jen changed my life forever from the moment I saw her across the room at the White House staff mess in 1994 and she has believed in me always and believed in this project from the beginning and through the inevitable ups and downs of writing. And our boys have grown up since that day in 2009 when they sat in the first row in the Benjamin Franklin room to watch Hillary Clinton formally welcome me to the State Department. I have had no greater joy, over the years, than being there for them and watching them blossom and grow. My brothers, Jeff and Jim, have just always been there with me and for me from the beginning. My sisters-in-law, Robin and Kathryn have been wonderful friends for decades. My mother- and father-in-law Carol and Allen Klein, have been supportive of my project throughout and, more important, supportive of our family. And of course I am deeply grateful to my late mother and father, who believed in me implicitly, always, and gave me the chance to find my calling.

Glossary

Annex 1; non-Annex 1: Annex 1 refers to the countries listed in Annex 1 to the 1992 UN Framework Convention on Climate Change and are essentially developed countries. Non-Annex 1 are the countries not listed in Annex 1 and are considered developing countries. Although these categories are more than thirty years old, they have never changed, largely owing to the politics of the UNFCCC.

AOSIS: The Alliance of Small Island States is an intergovernmental organization of low-lying coastal and small island countries.

BASIC: The group made up of China, India, Brazil, and South Africa, was formed in late 2009.

CBDR/RC and CBDR: "Common but differentiated responsibilities and respective capabilities," the phrase from the Framework Convention that established the principle of differentiation as between Annex 1 countries (essentially, developed countries) and non-Annex 1 countries (developing countries).

Celsius and Fahrenheit: Translating Celsius to Fahrenheit: C = (°F – 32) × 5/9. Translating Fahrenheit to Celsius: F = (°C × [9/5] + 32).

CCS: Carbon Capture and Storage technology.

CCUS: Carbon Capture, Utilization and Storage technology—in which captured carbon is not simply stored, geologically, but is put to some kind of constructive use.

COP: Conference of the Parties; the annual meeting of all parties to the UNFCCC, held late in the year.

GDP: Gross domestic product.

GT: Gigatons, equal to a billion tons, used in the climate change context as a means of measuring emissions.

G77 (or G77 + China): The Group of 77 developing countries founded in 1964, now numbering 135 countries. China considers itself a developing country but does not consider itself a member of the G77, so the group is often referred to as the G77 + China.

HFCs: Hydrofluorocarbons, an industrial greenhouse gas originally developed to replace an ozone-depleting gas. HFCs are powerful though short-lived climate pollutants.

HAC: The High Ambition Coalition, an important group of developed and developing countries that came together at the Paris COP to advocate for a strong, high-ambition Paris Agreement.

IEA: The International Energy Agency.

Indaba: Among the Bantu peoples of southern Africa, a meeting to discuss a serious topic. The term was first used in the UNFCCC context at the Durban (South Africa) COP in 2011 and then used again at the Paris COP in 2015.

Intersessional: A formal meeting to prepare for the COP. Usually there are two, three, or even four intersessionals, depending on the year. Typically they are attended by senior negotiators, not ministers.

Kyoto Protocol: Agreed to at the Kyoto COP in 1997, it was the first operational agreement intended to carry out the principles of the Framework Convention.

LDC: The Least Developed Country, a group of 45 poor countries, mostly from Africa and Asia. Countries can graduate out of the LDC group on the basis of economic development and resilience. Seven countries have graduated out of the LDC group, most recently Bhutan, in December 2023.

LMDC: The Like-Minded Developing Country group, founded in 2012, is made up of countries, including China and India, who are strong supporters of a sharp, categorical distinction between what developed countries are supposed to do and what developing countries are supposed to do. There are roughly 20–25 members, but membership has shifted over time.

MDBs: Multilateral development banks, which include the World Bank, Inter-American Development Bank, Asian Development Bank, the African Development Bank, and the European Bank for Reconstruction and Development.

MEF: The Major Economies Forum on Energy and Climate, typically referred to as the Major Economies Forum or MEF. A group of leading developed and developing countries founded and led by the United States.

MRV/transparency: Measurement, reporting, and verification, more commonly referred to as transparency, requiring countries to track their greenhouse gas inventories as well as their progress in carrying out their emission targets, prepare a report, and have the report reviewed by experts.

NDC/INDC: Nationally determined contributions or intended nationally determined contributions. An NDC is basically the emission target that a country submits periodically to the UNFCCC secretariat. The INDC is supposed to be submitted a number of months before the COP in the year that a new NDC is due so as to give others an opportunity to review and analyze the INDC.

NGOs: Nongovernmental organizations.

OECD: The Organization for Economic Cooperation and Development. An organization founded in 1961 that largely includes developed countries, but which a number of countries in the developing country group ("non-Annex 1") in the 1992 Framework Convention have joined, including Chile, Colombia, Costa Rica, Israel, Korea, Mexico.

Plenary: A plenary session is a meeting of all country parties to the UNFCCC or Paris Agreement, whether at a COP or intersessional meeting.

Pre-COP: A pre-COP is a meeting generally hosted by the host of the COP itself held a few weeks before the main conference to help prepare for the conference.

Secretariat: The UNFCCC secretariat is the United Nations entity tasked with supporting the global response to the threat of climate change. Among many other things, it helps organize and support all the major formal meetings each year, including the COP, and helps manage the activities of numerous initiatives and groups that have been and continue to be created by the Conference of the Parties.

SIDS: Small island developing states, a group of low-lying island nations that are home to approximately sixty-five million people and extremely vulnerable to the impacts of climate change.

Support: Generally means financial support provided to developing countries.

Umbrella Group: The United States and a number of other non-EU developed countries, including Japan, Canada, Australia, New Zealand, Norway, Ukraine, Iceland, Israel, and Kazakhstan. In 2023, the United Kingdom formally joined the group, which was first formed at Kyoto in 1997.

UNFCCC: The UN Framework Convention on Climate Change, agreed to in 1992. The foundational climate change agreement.

Notes

Prologue

1. The staff secretary controls the flow of all paper to the president, including decision memos, speeches, legislation to sign, and so on, decides what other senior staff need to comment, and in general plays an important role in the decision-making process in the White House.

2. The UNFCCC refers both to the original climate change treaty itself and to the UN body which, with its large Secretariat, oversees the international climate change regime.

Introduction

1. Joby Warrick and Chris Mooney, "196 Countries Approve Historic Climate Agreement," *Washington Post,* December 12, 2015, https://www.washingtonpost.com/news/energy-environment/wp/2015/12/12/proposed-historic-climate-pact-nears-final-vote.

2. Suzanne Goldenberg et al., "Paris Climate Deal: Nearly 200 Nations Sign in End of Fossil Fuel Era," *Guardian,* December 12, 2015, https://www.theguardian.com/environment/2015/dec/12/paris-climate-deal-200-nations-sign-finish-fossil-fuel-era.

3. Coral Davenport, "Nations Approve Landmark Climate Accord in Paris," *New York Times,* December 12, 2015, https://www.nytimes.com/2015/12/13/world/europe/climate-change-accord-paris.html.

4. "Historical GHG Emissions," Climate Watch, 2022, https://www.climatewatchdata.org/ghg-emissions?breakBy=regions&end_year=2020®ions=WORLD%2COECD&start_year=1990.

5. Thorfinn Stainforth and Bartosz Brzezinski, "More than Half of All CO_2 Emissions since 1751 Emitted in the Last 30 Years," Institute for European Environmental Policy, April 29, 2020, https://ieep.eu/news/more-than-half-of-all-co2-emissions-since-1751-emitted-in-the-last-30-years.

6. Daniel Bodansky, Jutta Brunnée, and Lavanya Rajamani, *International Climate Change Law* (New York: Oxford University Press, 2017), 108.

7. "The adoption rate of innovations is nonlinear; it is slow at first, then rapidly rises before flattening out again as it reaches market saturation. Such trajectories of growth are commonly known as S-curve. The rapidly rising part of the S-curve is often underestimated in projections and expectations of new technologies. This is exactly what has happened with wind, batteries, and solar technologies in the past decade, with prices dropping faster and further than many believed possible." Laurens Speelman and Yuki Numata, *A Theory of Rapid Transition: How S-Curves Work and What We Can Do to Accelerate Them* (Basalt, CO: RMI, October 2022), 1, https://rmi.org/insight/harnessing-the-power-of-s-curves.

8. John D. Sutter, Joshua Berlinger, and Ralph Ellis, "Obama: Climate Agreement 'Best Chance We Have' to Save the Planet," CNN, December 14, 2015, https://www.cnn.com/2015/12/12/world/global-climate-change-conference-vote/index.html.

Chapter 1

1. Years later, looking back at the way I tended to approach negotiations, I thought of a line from a 2016 *Atlantic* interview with Henry Kissinger. Kissinger was talking about issues of war and peace, far from the terrain of climate negotiations, but his formula has much broader application: "The fundamental strategic question is: What is it that we will not permit. . . . And a second question is: What are we trying to achieve." Quoted in Jeffrey Goldberg, "The Lessons of Henry Kissinger," *Atlantic*, November 10, 2016, https://www.theatlantic.com/magazine/archive/2016/12/the-lessons-of-henry-kissinger/505868.

2. Office of the Press Secretary, "FACT SHEET: The Recovery Act Made the Largest Single Investment in Clean Energy in History, Driving the Deployment of Clean Energy, Promoting Energy Efficiency, and Supporting Manufacturing," White House, February 25, 2016, https://obamawhitehouse.archives.gov/the-press-office/2016/02/25/fact-sheet-recovery-act-made-largest-single-investment-clean-energy.

3. Eamon Javers and Mike Allen, "Obama Announces New Fuel Standards," Politico, May 19, 2009, https://www.politico.com/story/2009/05/obama-announces-new-fuel-standards-022650.

4. "New Mileage Standards for Less Fuel, Pollution," CBS News, July 29, 2011, https://www.cbsnews.com/news/new-mileage-standards-for-less-fuel-pollution.

5. Jay Yarow, "Obama Asks For Cap And Trade Legislation," Business Insider, February 24, 2009, https://www.businessinsider.com/obama-asks-for-cap-and-trade-legislation-2009-2.

6. In 2009, transparency measures were still largely referred to as measurement, reporting, and verification (MRV). This started changing later in the year,

though both terms were sometimes used in the years leading up to the Paris COP in 2015.

7. Toyako Hokkaido, "Declaration of Leaders Meeting of Major Economies on Energy Security and Climate Change," G7 Research Group - University of Toronto, July 9, 2008, http://www.g8.utoronto.ca/summit/2008hokkaido/2008-mem.html. The MEM countries were the United States, United Kingdom, France, Germany, Italy, European Union, Canada, Japan, Australia, Russia, China, India, Brazil, South Africa, Korea, Indonesia, and Mexico.

8. Antholis was then at the Brookings Institution. We envisioned the following countries or entities in an E8: the United States, European Union, Japan, Russia, China, India, Brazil, and South Africa. Todd Stern and William J. Antholis, "Climate Change: Creating an E8," Brookings Institution, January 1, 2007, https://www.brookings.edu/articles/climate-change-creating-an-e8.

9. See "President Bush Discusses Climate Change," White House, April 16, 2008, https://georgewbush-whitehouse.archives.gov/news/releases/2008/04/20080416-6.html.

10. In related contexts, such as the G7 or G20, "leaders representatives" are often referred to as "Sherpas."

11. The fall of the Berlin Wall and integration of former Iron Curtain countries with high emissions into the European Union meant that EU emissions spiked up high by 1990, making that year a particularly favorable baseline for the European Union against which to measure its emission reductions after 1990.

12. Mark Stevenson, "Mexico: 'Green Fund' Better Than Carbon Credits," *San Diego Union-Tribune*, June 22, 2009, https://www.sandiegouniontribune.com/sdut-lt-mexico-climate-forum-062209-2009jun22-story.html.

13. "Fact Sheet: President Biden to Catalyze Global Climate Action through the Major Economies Forum on Energy and Climate," The White House, April 20, 2023, https://www.whitehouse.gov/briefing-room/statements-releases/2023/04/20/fact-sheet-president-biden-to-catalyze-global-climate-action-through-the-major-economies-forum-on-energy-and-climate.

14. Italian prime minister Silvio Berlusconi shifted the site of the G8 to L'Aquila out of respect for the people of Abruzzi (northeast of Rome), which had been struck by an earthquake just a few months earlier. "New Quake Hits G8 Summit Venue in Italy," *Sydney Morning Herald*, July 3, 2009, https://www.smh.com.au/world/new-quake-hits-g8-summit-venue-in-italy-20090703-d7ws.html.

15. Gandhi said at that time, "We do not wish to impoverish the environment any further and yet we cannot for a moment forget the grim poverty of large numbers of people." "Are not poverty and need the greatest polluters?" See Karl Mathiesen,

"Climate Change and Poverty: Why Indira Gandhi's Speech Matters," *Guardian*, May 6, 2014, https://www.theguardian.com/global-development-professionals-network/2014/may/06/indira-gandhi-india-climate-change.

16. Quoted in Matthew Rosenberg, "India Rejects U.S. Proposal of Carbon Limits," *Wall Street Journal*, July 20, 2009, https://www.wsj.com/articles/SB124789530843561651. We were not pressing India to reduce emissions since "reduce" in this usage meant an absolute reduction, as distinguished from a sharp lowering of emissions growth, which would be referred to as a "limit."

Chapter 2

1. MRV and transparency are essentially the same thing, involving both a country's reporting on its own efforts and an international review of those efforts. As we will see, "MRV" had connotations that developing countries did not like, so at a certain point, we started talking about "transparency" instead.

2. Climate change was a highly politicized issue in Australia, a bit like in the United States, but its government in 2009, led by Kevin Rudd, was a strong supporter of climate action.

3. There was already a detailed system of reporting for developed countries, including visits by expert review teams.

4. The G77 is the UN group of developing countries. Founded in 1964 with 77 developing countries, it now numbers 134. The G77 lists China as a member. Although the Chinese government has provided consistent political and financial support, and official statements include China, it does not consider itself to be a member. Hence the group is often referred to as the G77 + China.

5. Ananth Krishnan, "India, China Ready Climate Draft," *Hindu*, November 28, 2009, https://www.thehindu.com/news/international/India-China-ready-climate-draft/article16894829.ece.

6. The name comes from the initials of the four countries in Portuguese: Brazil, South Africa, India, and China.

7. The countries included China, India, Brazil, South Africa, Sudan for the G77, Korea, Ethiopia, Grenada, Algeria, Barbados, Bangladesh, the European Union, United Kingdom, Germany, France, and Australia, among others.

Chapter 3

1. John Vidal, "Copenhagen Climate Summit in Disarray after 'Danish Text' Leak," *Guardian*, December 8, 2009, https://www.theguardian.com/environment/2009/dec/08/copenhagen-climate-summit-disarray-danish-text.

2. The ministerial portion of the COP starts in the second week of these two-week meetings. I always made a point of arriving earlier than most ministers, and for the Copenhagen conference came particularly early. But ministers rarely arrive at the beginning.

3. "China Says U.S. Envoy 'Irresponsible' on Climate Aid," Reuters, December 11, 2009, https://www.reuters.com/article/us-climate-copenhagen-china-idUSTRE5BA3PM20091211.

4. To compare China and US emissions over time, see Hannah Ritchie, Pablo Rosado, and Max Roser, "Greenhouse Gas Emissions," Our World in Data, June 10, 2020, https://ourworldindata.org/greenhouse-gas-emissions.

5. In 2012, progressive Latin countries such as Chile, Colombia, Peru, and Costa Rica established the Independent Alliance of Latin America and the Caribbean.

6. Friends of the Chair is a name sometimes used for a procedure that the chair of a large negotiation employs to get a manageable number of parties into a room to hash out compromises and get a deal done.

7. Long-term cooperative action (LCA) was the negotiating track established at the 2008 Bali COP seeking a new agreement for all parties, as distinguished from the Kyoto negotiating track.

8. Interestingly, when Brown discussed his idea with the MEF countries at the London MEF in October, he ignored the firewall by describing China as well as the United States, European Union, and Japan as the biggest funders of the $100 billion.

9. The difference between countries providing versus mobilizing funds is important. "Provide" is understood to mean public money from government coffers. "Mobilize" is understood to mean both public money and private funds mobilized by government actions like loan guarantees, political risk insurance, and various other financial tools.

10. "Stand behind" was a phrase I had started to use in the fall MEF meetings because developing countries would not countenance the notion that they would "commit" to implement their pledged targets/actions, since they heard the word "commit" as suggesting legal bindingness and applying only to developed countries. To avoid that conundrum but still get our point across, I said that surely they would be prepared to stand behind what they promised. Dressed up in this colloquial manner, they went along.

11. Rasmussen also announced that the proposed decision to adopt the Copenhagen Accord would be accompanied by short decisions extending the efforts of the Kyoto and long-term cooperative action working groups for another year to pursue a new commitment period for Kyoto parties and a legally binding long-term cooperative action treaty.

12. The ALBA group—formally the Bolivarian Alliance for the Peoples of Our America—was launched in 2004 by Hugo Chávez of Venezuela and Cuba's Fidel Castro. By 2009, it included Bolivia, Ecuador, and Nicaragua, among others. All of these players were generally anti-Western and anti–United States.

13. In the United Nations' Glossary of Climate Change Acronyms and Terms, MISC "denotes a Miscellaneous document. These documents are not translated and are issued on plain paper with no United Nations masthead. In the UNFCCC process, submissions by Parties are normally issued as miscellaneous documents. They generally contain views or comments published as received from a delegation without formal editing."

14. See Yvonne Bell, "Copenhagen Accord Was a 'Disaster,' Says Sweden," Reuters, December 22, 2009, https://www.reuters.com/article/uk-climate-eu-copenhagen-idUKTRE5BL2WT20091222.

15. In reality, Obama was being ushered onto his plane so he could arrive in Washington before the blizzard that was bearing down on the city, and it was quite normal that he should say a few words to the traveling press before his departure.

16. John Vidal, "Ed Miliband: China Tried to Hijack Copenhagen Climate Deal," *Guardian*, December 20, 2009, Environment, https://www.theguardian.com/environment/2009/dec/20/ed-miliband-china-copenhagen-summit.

17. Ed Miliband, "The Road from Copenhagen," *Guardian*, December 20, 2009, https://www.theguardian.com/commentisfree/2009/dec/20/copenhagen-climate-change-accord.

18. See, for example, Michael Levi, "Beyond Copenhagen: Why Less May Be More in Global Climate Talks," *Foreign Affairs*, February 22, 2010, https://www.foreignaffairs.com/articles/commons/2010-02-22/beyond-copenhagen; Daniel Bodansky, "The Copenhagen Climate Change Conference: A Postmortem," *American Journal of International Law* 104, no. 2 (April 2010): 230–240, https://doi.org/10.5305/amerjintelaw.104.2.0230; Elliot Diringer, "Designing a Deal," Center for Climate and Energy Solutions, May 2010, https://www.c2es.org/document/designing-a-deal.

19. The word used in the Copenhagen Accord for developing country measures to limit emissions is "actions" while the parallel word used for developed countries is "targets." But in reality, the actions submitted by major developing countries, including all four BASIC countries, were numerical targets.

20. The accord treats supported and unsupported differently because "supported actions" are subject to "international MRV" on the theory that this would be more rigorous than international consultations and analysis. But the important conceptual breakthrough was that all actions were to be subject to some form of international review. And as a practical matter, a separate system of international MRV for supported actions never developed.

21. Our European friends sometimes saw us as obsessed about the firewall yet not firm enough about ambition, but I never bought that argument. Undermining the firewall *was* proambition in a world where China was becoming the dominant emitter and developing country emissions overall outpaced developed country emissions by a larger and larger margin with every passing year.

22. See, for example, Mark Lynas, "How Do I Know China Wrecked the Copenhagen Deal? I Was in the Room," *Guardian*, December 22, 2009, https://www.theguardian.com/environment/2009/dec/22/copenhagen-climate-change-mark-lynas; Vidal, "Ed Miliband."

Chapter 4

1. By March 2010, 73 parties, including 33 developing countries and all the majors, submitted their targets. By the end of 2010, some 140 countries had associated with the accord. Jacob Werksman, "Nearly 100 Countries Formally 'Associate' with Copenhagen Accord," *Inside Climate News,* March 25, 2010, https://insideclimatenews.org/news/25032010/nearly-100-countries-formally-associate-copenhagen-accord; "Summary of the Cancun Climate Change Conference: 29 November–11 December 2010," *Earth Negotiations Bulletin* 12, no. 498 (December 13, 2010), 2–3, https://s3.us-west-2.amazonaws.com/enb.iisd.org/archive/download/pdf/enb12498e.pdf?X-Amz-Content-Sha256=UNSIGNED-PAYLOAD&X-Amz-Algorithm=AWS4-HMAC-SHA256&X-Amz-Credential=AKIA6QW3YWTJ6YORWEEL%2F20240208%2Fus-west-2%2Fs3%2Faws4_request&X-Amz-Date=20240208T183056Z&X-Amz-SignedHeaders=host&X-Amz-Expires=60&X-Amz-Signature=7f1566ba027d8c863b0160494ac0db2e3180091a6a6662f2adf19e97643b6b77.

2. The Cartagena membership wasn't fixed, but the following countries attended the third meeting in 2010: Antigua and Barbuda, Australia, Bangladesh, Belgium, Chile, Colombia, Costa Rica, Denmark, Dominican Republic, Ethiopia, France, Germany, Guatemala, Indonesia, Malawi, Maldives, Marshall Islands, Mexico (as president for COP 16), Netherlands, New Zealand, Norway, Panama, Peru, Rwanda, Samoa, Spain, and the United Kingdom. "Third Meeting of the Cartagena Dialogue for Progressive Action, 31 Oct - 2 Nov. 2010, San Jose, Costa Rica," accessed February 5, 2024, https://ecbi.org/sites/default/files/Cartagenadialogue.pdf.

3. The presidency of the COP changes every year, chosen on a rotating basis by one of the United Nations' five regional groupings: Africa, Asia Pacific, eastern Europe, Latin America, and western Europe and others (among which "others" is the United States).

4. John M. Broder, "Obama to Face New Foes in Global Warming Fight," *New York Times,* November 3, 2010, https://www.nytimes.com/2010/11/04/business/energy-environment/04enviro.html.

5. Ramesh referred to his bullet points as an international consultations and analysis proposal, but in fact it was a proposal for a full transparency system. Technically, international consultations and analysis referred to the review piece of an overall system that started with a country submitting a transparency report including a number of specified elements. But Ramesh seized international consultations and analysis as the phrase du jour and used it for the title of his overall system.

6. The BASIC group put out an "experts" report in December 2011 showing not only that developed countries had *no* carbon space left but also that they had huge carbon deficits, which could only be addressed by providing enormous sums of money to developing countries.

7. The Conference of the Parties has actually never adopted a rule of voting, but has operated as a matter of general agreement on the basis that decisions would be made by consensus. Daniel Bodansky, Jutta Brunnée, and Lavanya Rajamani, *International Climate Change Law* (New York: Oxford University Press, 2017), 108.

8. Lisa Friedman, "A Near-Consensus Decision Keeps U.N. Climate Process Alive and Moving Ahead," *New York Times,* December 13, 2010, https://archive.nytimes.com/www.nytimes.com/cwire/2010/12/13/13climatewire-a-near-consensus-decision-keeps-un-climate-p-77618.html.

Chapter 5

1. Steinberg made his comments about Cancún at the Center for Strategic and International Studies on May 9, 2011, in response to a question about work the Arctic Council was trying to do on short-lived greenhouse gases, which are potent but linger in the atmosphere for much less time than CO_2.

2. Article II, Section 2, Clause 2 grants the president the power to "make treaties . . . by and with the advice and consent of the Senate," stipulating that two-thirds of the senators present must concur. U.S. Const., Art. II, § 2, Cl. 2.

3. The 1992 UNFCCC, concluded in the last year of the Bush administration, did require the "advice and consent" of the Senate, and was approved with little opposition, presumably because it imposed only general, not specific obligations on countries and was negotiated by a Republican administration. Five years later, after the Kyoto Protocol was agreed to at the 1997 COP, including legally binding emission targets for developed countries as well as a panoply of binding rules, the calculus was radically different, and the Clinton administration, whose president and vice president were very committed supporters of climate action, never sent the agreement to the Senate because it would have been dead on arrival.

4. Black helicopters refer to the specter of UN or shadowy new world order conspiracies out to undermine US sovereignty. See, for example, Lisa Caruso, "The 'Black

Helicopter Caucus,'" *Government Executive*, October 13, 1997, https://www.govexec.com/federal-news/1997/10/the-black-helicopter-caucus/4542.

5. In 2011, the European Union still had twenty-eight member states. Since the United Kingdom left after the Brexit vote, there have been twenty-seven member states.

6. There were a variety of potential escape hatch phrases, including "CBDR," a reference to the Bali Action Plan, and a reference to Article 4.7 of the Framework Convention, read by some developing countries (incorrectly in our view) to mean that any pledges they make to take action are conditional on the receipt of financial or technology support for developed countries.

7. Of course, the Kyoto Protocol was applicable to all parties in the purely formal sense that all COP parties had agreed to it and virtually all had formally joined it. But its only real obligations applied, as we have seen, to developed countries.

8. Lisa Friedman and Jean Chemnick, "Durban Talks Create 'Platform' for New Climate Treaty That Could Include All Nations," *E&E News*, December 12, 2011, https://subscriber.politicopro.com/article/eenews/1059957503.

Chapter 6

1. The Ad Hoc Working Group on the Durban Platform for Enhanced Action was the formal name of the negotiating track established in Durban. "Summary of the Bonn Climate Change Conference: 14–25 May 2012," *Earth Negotiations Bulletin* 12, no. 546 (May 28, 2012): 24, https://s3.us-west-2.amazonaws.com/enb.iisd.org/archive/download/pdf/enb12546e.pdf?X-Amz-Content-Sha256=UNSIGNED-PAYLOAD&X-Amz-Algorithm=AWS4-HMAC-SHA256&X-Amz-Credential=AKIA6QW3YWTJ6YORWEEL%2F20231030%2Fus-west-2%2Fs3%2Faws4_request&X-Amz-Date=20231030T013540Z&X-Amz-SignedHeaders=host&X-Amz-Expires=60&X-Amz-Signature=2900f6b0c5e26d4f0d226cbd670557707af364a0c6da74b4c880eea71bf732d3.

2. "Summary of the Bonn Climate Change Conference," 24–25.

3. I've long joked that track 2 meetings with China tend to be track 2 for the United States side and track 1.1 for China since its nongovernment participants are rarely more than a half step removed from the government.

4. The other, of course, is Fenway Park, home of the Boston Red Sox.

5. This was a process started by President George W. Bush's Treasury secretary, Hank Paulson, in 2007. Under Paulson, it was called the Strategic Economic Dialogue, met twice a year, and ran entirely out of the Treasury Department. When the Obama administration took office, Secretary Hillary Clinton at State and Secretary Timothy Geithner at Treasury modified the arrangements so that there were two separate

tracks, an economic track that Treasury would lead and a strategic track that State would lead—thus the slightly modified moniker, Strategic and Economic Dialogue.

6. The other two initial areas of cooperation were carbon capture, utilization, and storage along with collecting and managing greenhouse gas emissions data. Carbon capture and storage technology allows the carbon emissions from burning fossil fuels to be captured and sequestered underground. The technology is expensive, and the Chinese have always been keen to understand ways that the carbon could be used rather than buried since that would potentially be a more economically appealing proposition. Three more areas of cooperation were launched in 2014: forests, low-carbon cities, and industrial boiler efficiency / fuel switching. In 2015, an initiative on green ports and vessels was added.

7. See Melanie Hart, "China's Shifting Stance on Hydrofluorocarbons," Center for American Progress, June 12, 2013, https://www.americanprogress.org/article/chinas-shifting-stance-on-hydrofluorocarbons.

8. "The Kigali Amendment calls for a gradual reduction in the consumption and production of hydrofluorocarbons ('HFCs'), which are potent greenhouse gases. Its global implementation should avoid as much as half a degree Celsius of warming by the end of the century." "U.S. Ratification of the Kigali Amendment," US Department of State, September 21, 2022, https://www.state.gov/u-s-ratification-of-the-kigali-amendment.

9. For example, in the United Nations system, "countries with less than $1,035 Gross National Income per capita are classified as low-income countries, those with between $1,036 and $4,085 as lower middle income countries, those with between $4,086 and $12,615 as upper middle income countries, and those with incomes of more than $12,615 as high-income countries." *World Economic Situation and Prospects 2014* (New York: United Nations, 2014), 144, https://desapublications.un.org/file/519/download.

10. The phenomenon of huddles itself was striking in Warsaw. There is something quite different about the organic feel of a huddle as compared to sitting around a formal table. In a formal meeting, negotiators are likely more inclined to resist making concessions, and people generally only speak in an order that reflects when they raise their placard so that there is not much feel of give-and-take. A huddle tends to form because a few people start talking informally and in a less guarded manner. Ideas start popping out. Someone says, What about "initial" NDCs? Someone else doesn't like it and proposes something else. Ideas bounce around, and with ten or twenty self-selected people crowding in and hopefully a wise person acting as a makeshift moderator in the middle, if you're lucky, a huddle can produce a solution.

11. In fact, the Bali text said "actions" for developing countries and "commitments or actions" for developed, but after the Bali conference, developing countries generally airbrushed out "or actions" so that the takeaway was a clean distinction—actions

versus commitments—between developing countries and developed. There is, of course, no tribunal of wise ones to which disagreeing countries can take their language dispute, so what language ends up meaning tends to get settled in the "street" as much as anywhere else.

12. Lisa Friedman, "Is America No Longer Public Enemy No. 1 on Climate Change?," *E&E News*, November 21, 2013, https://www.eenews.net/stories/1059990838.

13. Center for Climate and Energy Solutions, "Outcomes of the U.N. Climate Change Conference in Warsaw," Center for Climate and Energy Solutions, November 2013, https://www.c2es.org/document/outcomes-of-the-u-n-climate-change-conference-in-warsaw.

14. Fiona Harvey, "As the Warsaw Climate Talks End, the Hard Work Is Just Beginning," *Guardian*, November 25, 2013, https://www.theguardian.com/environment/2013/nov/25/warsaw-climate-talks-end-cop19-2015.

Chapter 7

1. It was by no means clear to China specialists at the time whether Xi actually hoped for beneficent mutual relations going forward or just wanted to gull Americans into thinking so. Certainly some feared that Xi's slogan actually meant little more than that he wanted to see passive US responses to all manner of aggressive Chinese behavior. This was a matter of debate. But either way, Xi's emphasis on this formulation suggested that he might well decide it was worth working with the United States on this new, collaborative idea.

2. Erica was not only highly competent and effective in general but also played an important role in an impressive gambit at the US embassy in China that had significant impact for the whole country. In 2010, she took over management of an air quality monitoring program that published data from a PM2.5 monitor affixed to the US embassy roof. (Fine particulate matter, a dangerous form of pollution, is defined as particles that measure 2.5 microns or less.) Begun to keep US citizens informed about Beijing's often treacherous air pollution, the program's information started to spread to ordinary Beijing residents despite the efforts of Beijing authorities. Those residents previously had no accurate data, and became more and more interested in and concerned about the data from the US embassy. Eventually, widespread public pressure led the Chinese central government to issue new air pollution guidelines in November 2011, including, for the first time, standards for PM2.5. From then on, the Chinese central government made the issue of air quality a centerpiece of its environmental strategy.

3. Mike was appointed by the president to be the US trade representative.

4. Zhang was later accused of sexual assault by tennis player Peng Shuai. Nectar Gan and Steve George, "Who Is Zhang Gaoli? The Man at the Center of Chinese Tennis

Star Peng Shuai's #MeToo Allegation," CNN, November 25, 2021, https://www.cnn.com/2021/11/25/china/who-is-zhang-gaoli-intl-hnk-dst/index.html.

5. Editorial Board, "A Major Breakthrough on Climate Change," *New York Times*, November 12, 2014, https://www.nytimes.com/2014/11/13/opinion/climate-change-breakthrough-in-beijing.html.

6. Max Fisher, "Obama's Climate Deal Proves China Is the Biggest Foreign Policy Success of His Presidency," *Vox*, November 12, 2014, https://www.vox.com/2014/11/12/7203263/obama-china-climate.

Chapter 8

1. There were actually two versions of a text marked December 8 at 6:30 p.m. One was just six pages long plus an eleven-page annex on the information parties needed to provide with their NDCs. The other was thirty-three pages long and probably produced after the cochairs of the negotiation opened up the text for alternative language proposals on December 8.

2. "Submission by Greece and the European Commission on Behalf of the European Union and Its Member States," May 28, 2014, 4–5, https://unfccc.int/files/bodies/awg/application/pdf/el-05-28-adp_ws1_submission.pdf.

3. Simon Romero and Randal C. Archibold, "Brazil Angered over Report N.S.A. Spied on President," *New York Times*, September 3, 2013, https://www.nytimes.com/2013/09/03/world/americas/brazil-angered-over-report-nsa-spied-on-president.html.

4. A contact group is usually convened at the direction of the chair of a formal process, and can range in size from a small group of countries meeting to discuss and recommend solutions to a particular problem to all the parties of an overall negotiation. The contact group in Lima was of the second variety.

5. This sentence is an interesting one. The notion that many of you colonized us and that's why there must be differentiated treatment between developed and developing countries (Annex 1 and non-Annex 1) is not self-evident purely from an emissions standpoint. What it reflects is a reality often commented on, that other "isms"—such as imperialism or colonialism—representing unfair treatment, have at times gotten wrapped up in the anger of developing countries against developed.

6. Lisa Friedman, "After 30 Hours of Overtime Haggling, Diplomats Agree on a New Approach to Deal with Climate Change," *E&E News*, December 14, 2014, https://www.eenews.net/greenwire/stories/1060010528.

7. Quoted in Ed King and Megan Darby, "LIVE IN LIMA—DAY 12: UN COP20 Climate Change Summit," *Climate Home News*, December 13, 2014, https://www.climatechangenews.com/2014/12/13/live-in-lima-day-12-un-cop20-climate-change-summit.

8. Quoted in Ed King, “US-China Chat Broke Impasse at Lima Climate Talks,” *Climate Home News*, December 16, 2014, https://www.climatechangenews.com/2014/12/16/us-china-chat-broke-impasse-at-lima-climate-talks.

9. “Summary of the Lima Climate Change Conference: 1–14 December 2014,” *Earth Negotiations Bulletin* 12, no. 619 (December 16, 2014): 44, https://s3.us-west-2.amazonaws.com/enb.iisd.org/archive/download/pdf/enb12619e.pdf?X-Amz-Content-Sha256=UNSIGNED-PAYLOAD&X-Amz-Algorithm=AWS4-HMAC-SHA256&X-Amz-Credential=AKIA6QW3YWTJ6YORWEEL%2F20240205%2Fus-west-2%2Fs3%2Faws4_request&X-Amz-Date=20240205T234220Z&X-Amz-SignedHeaders=host&X-Amz-Expires=60&X-Amz-Signature=75e989bb4ef77a770cc52c2846546e3ca9b87a8a51d3364f4d31f69813e16d2e.

10. Michael Jacobs, “Lima Deal Represents a Fundamental Change in Global Climate Regime,” *Guardian*, December 15, 2014, https://www.theguardian.com/environment/2014/dec/15/lima-deal-represents-a-fundamental-change-in-global-climate-regime.

11. “The Guardian View on the Lima Climate Change Conference: A Skirmish before the Real Battle,” *Guardian*, December 14, 2014, https://www.theguardian.com/commentisfree/2014/dec/14/guardian-view-lima-climate-change-conference-cop-20-skirmish1.

Chapter 9

1. I got off to a mildly awkward start with Fabius at a dinner he hosted for climate ministers at the Château de La Celle Saint-Cloud during the July 2014 MEF meeting in Paris. He had thoughtfully seated me at his table to his immediate right. And although I was reasonably well-bred and even spoke some rusty French, the elegant, abundant china and cutlery left me unsure about which bread plate—left or right—belonged to me. After hesitating for a few minutes, I took a stab at it, and in a word, ate Fabius's roll—thus sadly playing the predictable part of the not entirely couth American. I expect he doesn't even remember this inauspicious debut. He even graciously pretended not to notice. I did better with him after that.

2. Xie and I had actually discussed the notion of resolving important issues before Paris back in March in Beijing during an odd, in-between period when Xie was still formally retired—a move the Chinese government announced not long after the Lima COP, owing to Xie having turned sixty-five. I was amazed that China would sideline its lead climate player, and in the end, it did not. But in March, not knowing what China's plan was, I traveled to Beijing to meet with the official designated to replace Xie, Zhang Yong, but I also had dinner with Xie, who was still completely on top of his climate brief. Less than a month later, he was unretired and back in Washington at our MEF meeting as a “special representative” leading China in the climate negotiations.

3. We were comfortable that a legally binding transparency system would not trigger the need for Senate approval because a new system could be fairly regarded as an extension of the transparency article in the Framework Convention, which had, of course, been approved by the Senate in 1992.

4. Brian Kahn, "G7 Leaders: World Needs to Phase Out Carbon Emissions | Climate Central," Climate Central, June 8, 2015, https://www.climatecentral.org/news/g7-carbon-emissions-19082.

5. Quoted in Julie Hirschfeld Davis, "For Obama and Indian Leader, a Friendly Stroll If Not a Full Embrace," *New York Times*, September 30, 2014, https://www.nytimes.com/2014/10/01/world/asia/for-obama-and-modi-meetings-dc-mlk-monument.html.

6. Geeta Anand, "India, Once a Coal Goliath, Is Fast Turning Green," *New York Times*, June 2, 2017, https://www.nytimes.com/2017/06/02/world/asia/india-coal-green-energy-climate.html.

7. I never thought, by the way, that political support in the United States only mattered if the agreement needed to be approved by the Senate. Political support would be critically important in any event to help drive domestic action to meet our target and to allow the United States to be an influential player in building a strong international regime.

8. The agreement was accompanied by an eleven-page draft decision, providing more operational detail and guidance about the next steps.

9. "Nozipho Mxakato-Diseko—La diva du climat ressuscite Mandela," *L'Echo*, 2015, https://www.lecho.be/dossier/paris2015/Nozipho_Mxakato_Diseko_La_diva_du_climat_ressuscite_Mandela/9708403.html.

10. This was no doubt a product of the US opposition to a fully legally binding agreement as well as the preference of the European Union to play a leading role in an important group without having the United States there, with our unavoidably big footprint. But two things were different now. We were getting down to the short strokes in the negotiations, plus President Obama's sharply increased climate activism in his second term had burnished our credibility and standing.

11. Saudi Arabia had argued for years that climate negotiations had to take into account the damage done to fossil-fueled economies by measures to mitigate climate change, and Khalid wanted to make sure that the progress he had made over the years in allowing countries to report on the impact of such response measures would not be lost.

12. "Leaders and High-Level Segment," United Nations Climate Change, October 29, 2015, https://unfccc.int/news/cop-21cmp-11-information-hub-leaders-and-high-level-segment.

13. The three funds were the Caribbean Catastrophe Risk Insurance Facility, Pacific Catastrophic Risk Assessment and Financing Initiative, and African Risk Capacity.

14. The idea of Paris giving us a chance is the way the president often spoke about it. He used that language in his statement from the White House after the deal got done. Some weeks after COP 21, the president came to meet my whole, extended team and take a photo with us, along with Secretary Kerry. While he was there, he signed the first page of a copy of the Paris Agreement, writing, "To the Paris Team—you've given future generations a fighting chance!"

15. We made one mistake in an otherwise great meeting—not inviting Prime Minister Sopoaga of Tuvalu. I learned years later that he had been quite miffed at not being included. We had been careful in deciding who to invite, but somehow missed Tuvalu—something I should have noticed, since I had met with Prime Minister Sopoaga in New York in September and knew he cared a great deal about climate change.

16. Karl Mathiesen and Fiona Harvey, "Climate Coalition Breaks Cover in Paris to Push for Binding and Ambitious Deal," *Guardian,* December 8, 2015, https://www.theguardian.com/environment/2015/dec/08/coalition-paris-push-for-binding-ambitious-climate-change-deal.

17. Megan Darby and Ed King, "COP21: US Joins 'Coalition of High Ambition' at Paris Talks," *Climate Home News,* December 9, 2015, https://www.climatechangenews.com/2015/12/09/cop21-live-climate-talks-intensify-in-paris.

18. A question related to the five-year individual cycle that did not get resolved in Paris was known as "common time frames"—that is, whether countries would all submit their new NDCs in the same year. In Paris, most countries had chosen targets that would cover a ten-year period, from 2020 (when the agreement was presumed to take effect) to 2030, but some, including the United States and Marshall Islands, had chosen targets that would cover a five-year period, to 2025. We had argued all year that five-year targets would be more ambitious because countries would be continually forced to assess what they could do for each five years. We thought that given the rapidly changing landscape of climate change, letting targets run for ten years was too long. The issue finally got resolved at the Glasgow COP in 2021, with a decision that encouraged, but did not force, parties to submit new NDCs every five years. United Nation Framework Convention on Climate Change, "Decision-/CMA.3: Common Time Frames for Nationally Determined Contributions Referred to in Article 4, Paragraph 10, of the Paris Agreement," UNFCCC, November 2021, https://unfccc.int/sites/default/files/resource/cma3_auv_3b_CTF.pdf.

19. Quoted in Megan Darby and Ed King, "As It Happened: Paris Climate Pact Edged Closer on Thursday," *Climate Home News*, December 10, 2015, https://www.climatechangenews.com/2015/12/10/cop21-live-clock-ticking-on-paris-climate-talks.

20. Energy Reporter, "Climate Deal 'Meaningless' without Five Year Reviews," *Energy Voice*, December 10, 2015, https://www.energyvoice.com/other-news/95499/climate-deal-meaningless-without-five-year-reviews.

21. That the French borrowed this South African word, "Indaba," popularized in the UNFCCC during COP 17 in Durban (2011), was another nice touch. It showed respect for the process South Africa inaugurated in 2011 and this South African word, which typically refers to a conference where important matters were discussed among the Zulu or Xosha peoples of South Africa.

22. This was another one of those times when the South African climate identity was so interesting. This was the South Africa—a member of BASIC—that could sometimes out–Europe the European Union, taking a position that in essence, was a version of the position the European Union pushed hard in Lima for an assessment process of country INDCs—known by the shorthand of *ex ante* and probably more hated by China than any other proposal.

23. The Chinese agreed to modify the traditional formula captured in the December 9 text of developed countries "taking the lead," often used to establish bifurcation—developed countries do x, developing may get there at some point, but not now. The new language said that developed countries "should" continue taking the lead by undertaking economy-wide absolute reduction targets (their standard form of targets), and that developing countries would be encouraged to move to these kinds of targets over time. The key points for us were that this was a nonlegally binding commitment for developed countries ("should," not "shall"), that developed countries were continuing to take the lead on a specific manner, not across the board, and that developing countries would be encouraged to do the same thing.

24. Alex Pashley, Megan Darby, and Ed King, "As It Happened: Paris COP21 Climate Talks Run into Overtime," *Climate Home News*, December 11, 2015, https://www.climatechangenews.com/2015/12/11/cop21-live-deadline-delayed-as-talks-get-tense.

25. Quoted in Alex Pashley, "Brazil Backs 'High Ambition Coalition' to Break Paris Deadlock," *Climate Home News*, December 11, 2015, https://www.climatechangenews.com/2015/12/11/brazil-backs-high-ambition-coalition-to-break-paris-deadlock.

26. Pashley, "Brazil Backs 'High Ambition Coalition' to Break Paris Deadlock."

27. Cited in Pashley, "Brazil Backs 'High Ambition Coalition' to Break Paris Deadlock."

28. Rémy Rioux, *Climate Finance: Reviving the Spirit of COP 21* (Paris: Agence Française de Développement, 2018), 10–11, https://www.afd.fr/sites/afd/files/2018-09-11-21-56/climate-finance-reviving-spirit-cop21-remy-rioux.pdf.

29. See, for example, Rioux, *Climate Finance*.

30. Net-zero emissions means that the emissions caused by human activity are balanced by the removal of emissions from the atmosphere, whether by natural means

such as trees that absorb emissions or technological measures that can remove emissions from the atmosphere. Kelly Levin, Taryn Fransen, Clea Schumer, Chantal Davis, and Sophie Boehm, "What Does 'Net-Zero Emissions' Mean? 8 Common Questions, Answered," World Resources Institute, March 20, 2023, https://www.wri.org/insights/net-zero-ghg-emissions-questions-answered.

31. Farhana, a fierce, implacable British lawyer and activist, first pitched her net-zero idea to me, Sue and my chief of staff Kareem Salah in July 2013 in London at Fortnum and Mason's, her guilty pleasure. When she laid out her idea of a net-zero goal for the world by 2050, it seemed so diplomatically outlandish at the time that I nearly laughed. But Farhana was dead serious. I agreed with her communication point that people don't get what 1.5 or 2 degrees means, but understand what "zero" means. But I couldn't imagine countries endorsing net zero by 2050 when they wouldn't accept even a global 50 percent cut by 2050 just a few years earlier. But Farhana turned out to be right: not about the politics of 2013, but about where we needed to go.

32. Sadly, Minister Molewa became ill on a trip to China in 2018 and died shortly after the trip.

33. See Dan Bodansky, "Reflections on the Paris Conference," *Opinio Juris* (blog), December 15, 2015, https://opiniojuris.org/2015/12/15/reflections-on-the-paris-conference.

34. Alex Pashley, Megan Darby, and Ed King, "As It Happened: 195 Countries Agree Paris Climate Deal," ," *Climate Home News*, December 12, 2015, https://www.climatechangenews.com/2015/12/12/195-countries-agree-paris-climate-deal.

35. Mohammed has been the deputy secretary-general of the United Nations since 2017.

36. Isabelle Gerretsen, "Climate Watchers Pay Tribute to Nicaraguan Envoy Paul Oquist, Who Died on Monday," *Climate Home News*, April 14, 2021, https://www.climatechangenews.com/2021/04/14/climate-watchers-pay-tribute-nicaraguan-envoy-paul-oquist-died-monday.

37. Roger Harrabin, "COP21: Did the Pope Save the Climate Deal?," *BBC News*, December 13, 2015, https://www.bbc.com/news/science-environment-35087220.

38. Oquist worked in the Sandinista governments of Ortega in the 1980s, eventually becoming a climate negotiator. He was an odd though interesting guy, with whom I once had a quite interesting dinner conversation when we found ourselves seated at the same table.

39. Quoted in Lire Aussi, "Une Dernière Journée Marathon Avant L'adoption d'un «Accord Décisif pour la Planète»," *Le Monde*, December 12, 2015, https://www.lemonde.fr/cop21/article/2015/12/12/cop21-laurent-fabius-presente-un-texte-d-accord-mondial-sur-le-climat_4830539_4527432.html.

40. Quoted in Pashley, Darby, and King, "As It Happened: 195 Countries Agree Paris Climate Deal."

41. Jonathan Chait, "The Paris Climate Deal Is President Obama's Biggest Accomplishment," *Intelligencer* December 14, 2015, https://nymag.com/intelligencer/2015/12/climate-deal-is-obamas-biggest-accomplishment.html.

42. Editorial Board, "The Paris Climate Pact Will Need Strong Follow-up," *New York Times*, December 14, 2015, https://www.nytimes.com/2015/12/15/opinion/the-paris-climate-pact-will-need-strong-follow-up.html.

43. See, for example, Fiona Harvey, "Paris Climate Change Deal Too Weak to Help Poor, Critics Warn," *Guardian*, December 14, 2015, Environment, https://www.theguardian.com/environment/2015/dec/14/paris-climate-change-deal-cop21-oxfam-actionaid; Tom Bawden, "Paris Climate Deal 'Far Too Weak to Prevent Devastating Global Warming,'" *Independent*, January 8, 2016, Climate, https://www.independent.co.uk/climate-change/news/cop21-paris-deal-far-too-weak-to-prevent-devastating-climate-change-academics-warn-a6803096.html.

44. Brian Deese, "Paris Isn't Burning: Why the Climate Agreement Will Survive Trump," *Foreign Affairs*, May 22, 2017, https://www.foreignaffairs.com/world/paris-isnt-burning.

45. Michael Jacobs, "High Pressure for Low Emissions: How Civil Society Created the Paris Climate Agreement," Institute for Public Policy Research, March 14, 2016, https://www.ippr.org/juncture-item/high-pressure-for-low-emissions-how-civil-society-created-the-paris-climate-agreement.

Chapter 10

1. *The Paris Effect: How the Climate Agreement Is Reshaping the Global Economy* (London: Systemiq, December 2020), 13, https://www.systemiq.earth/wp-content/uploads/2020/12/The-Paris-Effect_SYSTEMIQ_Full-Report_December-2020.pdf.

2. United Nations, "COP21—Frequently Asked Questions," 2015, https://www.un.org/sustainabledevelopment/wp-content/uploads/2015/10/COP21-FAQs.pdf; Janet Tobias and Tess Wiskel, "At COP 28, Global Consensus on How Climate Change Affects Health," *PBS NewsHour*, December 13, 2023, World, https://www.pbs.org/newshour/world/at-cop-28-global-consensus-on-how-climate-change-affects-health.

3. Copernicus, the European climate monitor, found the global average temperature to be 1.48 degrees above nineteenth-century levels, while the US agencies NOAA (the National Oceanic and Atmospheric Administration) and NASA measured warming up by 1.35 and 1.4, respectively. Editorial Board, "The 1.5-Degree Climate Goal Is out of Reach. Here's What to Do Now," *Washington Post*, January 18, 2024, https://www.washingtonpost.com/opinions/2024/01/18/climate-change-target-missing-global-action.

4. Associated Press, "Phoenix Hit 110 Degrees on 54 Days in 2023, Setting Another Heat Record," *PBS NewsHour*, September 10, 2023, https://www.pbs.org/newshour/nation/phoenix-hit-110-degrees-on-54-days-in-2023-setting-another-heat-record.

5. Catrin Einhorn and Elena Shao, "How Hot Is the Sea Off Florida Right Now? Think 90s Fahrenheit," *New York Times*, July 12, 2023, https://www.nytimes.com/2023/07/12/climate/florida-ocean-temperatures-reefs.html.

6. Dan Bilefsky, "What to Know about Canadian Wildfires and U.S. Air Quality," *New York Times*, June 30, 2023, https://www.nytimes.com/2023/06/28/world/canada/canada-wildfires-smoke-us-air-quality.html.

7. Sam Meredith, "South American Countries Gripped by Record-Breaking Heat—in the Middle of Winter," CNBC, August 4, 2023, https://www.cnbc.com/2023/08/04/south-america-argentina-and-chile-gripped-by-extreme-mid-winter-heat.html.

8. "2023 Libya Floods," Center for Disaster Philanthropy, January 16, 2024, https://disasterphilanthropy.org/disasters/2023-libya-floods.

9. Alvin Lin, "China's Climate Progress Ahead of COP27," NRDC (Natural Resources Defense Council), November 1, 2022, https://www.nrdc.org/bio/alvin-lin/chinas-climate-progress-ahead-cop27.

10. "Death Toll from Brutal Heat Wave Tops 1,000 in Spain and Portugal," AccuWeather, July 18, 2022, https://www.accuweather.com/en/weather-forecasts/death-toll-from-brutal-heat-wave-tops-1000-in-spain-and-portugal/1218084.

11. Joan Ballester, Marcos Quijal-Zamorano, Raúl Fernando Méndez Turrubiates, Ferran Pegenaute, François R. Herrmann, Jean Marie Robine, Xavier Basagaña, et al., "Heat-Related Mortality in Europe during the Summer of 2022," *Nature Medicine* 29, no. 7 (July 10, 2023): 1861, https://doi.org/10.1038/s41591-023-02419-z.

12. Scott K. Johnson, "Pakistan Hits 120°F as Climate Trends Drive Spring Heatwave," Ars Technica, May 23, 2022, https://arstechnica.com/science/2022/05/pakistan-hits-120f-as-climate-trends-drive-spring-heatwave.

13. Sara E. Pratt, "Devastating Floods in Pakistan," Earth Observatory, –NASA, accessed January 10, 2024, https://earthobservatory.nasa.gov/images/150279/devastating-floods-in-pakistan.

14. Emily Cassidy, "Worst Drought on Record Parches Horn of Africa," Earth Observatory, NASA, accessed January 10, 2024, https://earthobservatory.nasa.gov/images/150712/worst-drought-on-record-parches-horn-of-africa.

15. Tamara White, "Increasing Droughts and Floods on the African Continent," Brookings, October 8, 2021, https://www.brookings.edu/articles/increasing-droughts-and-floods-on-the-african-continent.

16. Joshua Partlow, "Officials Fear 'Complete Doomsday Scenario' for Drought-Stricken Colorado River," *Washington Post,* December 1, 2022, https://www.washingtonpost.com/climate-environment/2022/12/01/drought-colorado-river-lake-powell.

17. Helen Davidson, "China Drought Causes Yangtze to Dry Up, Sparking Shortage of Hydropower," *Guardian,* August 22, 2022, https://www.theguardian.com/world/2022/aug/22/china-drought-causes-yangtze-river-to-dry-up-sparking-shortage-of-hydropower.

18. Jon Henley, "Europe's Rivers Run Dry as Scientists Warn Drought Could Be Worst in 500 Years," *The Guardian,* August 13, 2022, https://www.theguardian.com/environment/2022/aug/13/europes-rivers-run-dry-as-scientists-warn-drought-could-be-worst-in-500-years.

19. *The Paris Effect*, 13.

20. Kingsmill Bond, Sam Butler-Sloss, Amory Lovins, Laurens Speelman, and Nigel Topping, *X-change: Electricity* (Basalt, CO: Rocky Mountain Institute, July 2023), 5–6, https://rmi.org/wp-content/uploads/dlm_uploads/2023/07/rmi_x_change_electricity_2023.pdf.

21. Kingsmill Bond, Sam Butler-Sloss, Daan Walter, Harry Benham, E. J. Klock-McCook, Dave Mullaney, Yuki Numata, et al., *X-change: Cars: The End of the ICE Age* (Basalt, CO: Rocky Mountain Institute, September 2023), 6, https://rmi.org/wp-content/uploads/dlm_uploads/2023/09/x_change_cars_report.pdf.

22. "EVs to Surpass Two-Thirds of Global Car Sales by 2030, Putting at Risk Nearly Half of Oil Demand, New Research Finds," Rocky Mountain Institute, September 14, 2023, https://rmi.org/press-release/evs-to-surpass-two-thirds-of-global-car-sales-by-2030-putting-at-risk-nearly-half-of-oil-demand-new-research-finds.

23. Rupert Way and J. Doyne Farmer, "A Fast Clean Energy Transition Would Save Trillions," Research Briefing, Institute for New Economic Thinking at the Oxford Martin School, October 11, 2023,),2, https://www.inet.ox.ac.uk/files/OMS-energy-transition-v7.pdf.

24. Sam Butler-Sloss and Kingsmill Bond, "The Eight Deadly Sins of Analyzing the Energy Transition," RMI, October 13, 2023, https://rmi.org/the-eight-deadly-sins-of-analyzing-the-energy-transition.

25. Canary Staff, "Six Reasons to Be Optimistic about the Energy Transition," Canary Media, December 28, 2023, https://www.canarymedia.com/articles/clean-energy/six-reasons-to-be-optimistic-about-the-energy-transition.

26. *Energy Efficiency—The Decade for Action: Ministerial Briefing* (Versailles: International Energy Agency, June 2023), 8, https://iea.blob.core.windows.net/assets/f475eb45-f14f-4a6f-9217-51eadef32763/EnergyEfficiency-TheDecadeforAction.pdf.

27. Alfredo Rivera, Shweta Movalia, Hannah Pitt, and Kate Larsen, "Global Greenhouse Gas Emissions: 1990–2020 and Preliminary 2021 Estimates," Rhodium Group, December 19, 2022, 4–5, https://rhg.com/research/global-greenhouse-gas-emissions-2021.

28. Geological hydrogen is hydrogen found in the earth rather than the kind in use now that needs to be made using electrolyzers. "Cruisin' for Nonconventional Fusion with ARPA-E," ARPA-E, March 22, 2021, http://arpa-e.energy.gov/news-and-media/blog-posts/cruisin-nonconventional-fusion-arpa-e; Communications and Publishing, "The Potential for Geologic Hydrogen for Next-Generation Energy," USGS, April 13, 2023, https://www.usgs.gov/news/featured-story/potential-geologic-hydrogen-next-generation-energy.

29. *Energy Efficiency 2023* (Paris: International Energy Agency, November 2023), 11, https://iea.blob.core.windows.net/assets/dfd9134f-12eb-4045-9789-9d6ab8d9fbf4/EnergyEfficiency2023.pdf.

30. "Energy Efficiency," IEA, accessed January 12, 2024, https://www.iea.org/energy-system/energy-efficiency-and-demand/energy-efficiency.

31. *Energy Efficiency 2022* (Paris: International Energy Agency, November 2022), 8, https://iea.blob.core.windows.net/assets/7741739e-8e7f-4afa-a77f-49dadd51cb52/EnergyEfficiency2022.pdf.

32. *Energy Efficiency 2023*, 9.

33. "At IEA Conference, 46 Governments Endorse Goal of Doubling Global Energy Efficiency Progress by 2030," IEA, June 9, 2023, https://www.iea.org/news/at-iea-conference-46-governments-endorse-goal-of-doubling-global-energy-efficiency-progress-by-2030.

34. "The German Feed-in Tariff," futurepolicy.org, accessed January 12, 2024, https://www.futurepolicy.org/climate-stability/renewable-energies/the-german-feed-in-tariff; Isabel Sutton, "Germany: Will the End of Feed-in Tariffs Mean the End of Citizens-as-Energy-Producers," Energy Post, June 3, 2021, https://energypost.eu/germany-will-the-end-of-feed-in-tariffs-mean-the-end-of-citizens-as-energy-producers.

35. Richard Kessler, "'We Need Trillions': Top US Clean Energy Adviser John Podesta Tells Private Sector to Step Up," Recharge, May 24, 2023, https://www.rechargenews.com/energy-transition/we-need-trillions-top-us-clean-energy-adviser-john-podesta-tells-private-sector-to-step-up/2-1-1455256.

36. Jennifer Harris, "No Country Can Solve Critical Mineral Shortages Alone," *Financial Times*, July 7, 2023, https://www.ft.com/content/394dca37-ac50-4380-9b03-4fdfcef2ff7c.

37. "Renewable Energy Law of the People's Republic of China," IEA, April 16, 2021, https://www.iea.org/policies/3080-renewable-energy-law-of-the-peoples-republic-of-china.

38. "COP26: India PM Narendra Modi Pledges Net Zero by 2070," BBC, November 2, 2021, https://www.bbc.com/news/world-asia-india-59125143.

39. M. Crippa, D. Guizzardi, E. Schaaf, F. Monforti-Ferrario, R. Quadrelli, A. Risquez Martin, S. Rossi, et al., *GHG Emissions of All World Countries 2023* (Luxembourg: Joint Research Centre, European Commission, 2023), 5, https://data.europa.eu/doi/10.2760/953322.

40. Methane is short-lived, only lasting in the atmosphere for around twelve years, but it is potent, with some eighty times more warming power than CO_2. It is responsible for around 30 percent of global warming in the industrial era. The United States and European Union launched the Global Methane Pledge at the Glasgow COP in 2021, aimed at reducing methane emissions 30 percent by 2030. At the time of the Dubai COP, some 155 countries had signed on. "Methane and Climate Change—Global Methane Tracker 2022," IEA, February 2022, https://www.iea.org/reports/global-methane-tracker-2022/methane-and-climate-change; "What's the Deal with Methane?," UNEP, October 18, 2022, http://www.unep.org/news-and-stories/video/whats-deal-methane; "Homepage—Global Methane Pledge," Global Methane Pledge, accessed January 14, 2024, https://www.globalmethanepledge.org; Nick Robertson, "Oil and Gas Producers Pledge to Cut Methane Emissions at UN Climate Talks," *The Hill,* December 2, 2023, https://thehill.com/policy/energy-environment/4338870-oil-gas-producers-pledge-cut-methane-emissions-un-climate-talks; Bill Spindle, "Oil and Gas Companies Step into the Spotlight at COP28—Cipher News," *Cipher News*, December 6, 2023, https://ciphernews.com/articles/oil-and-gas-companies-step-into-the-spotlight-at-cop28.

41. Nick Robertson, "Oil and Gas Producers Pledge to Cut Methane Emissions at UN Climate Talks," *The Hill,* December 2, 2023, https://thehill.com/policy/energy-environment/4338870-oil-gas-producers-pledge-cut-methane-emissions-un-climate-talks.

42. "Statement: Oil and Gas Decarbonization Charter Unveiled," World Resources Institute, December 2, 2023, https://www.wri.org/news/statement-oil-and-gas-decarbonization-charter-unveiled.

43. Alvin Lin, "China Commits to Strictly Control and Phase Down Coal," NRDC, April 22, 2021, https://www.nrdc.org/bio/alvin-lin/china-commits-strictly-control-and-phase-down-coal.

44. John Kemp, "China's Rainfall Is in the Wrong Place for Hydropower," Reuters, August 23, 2023, https://www.reuters.com/business/energy/chinas-rainfall-is-wrong-place-hydropower-kemp-2023-08-22; "China Permits Two New Coal Power Plants per Week in 2022," CREA, February 2023), 2, https://energyandcleanair.org/wp/wp-content/uploads/2023/02/CREA_GEM_China-permits-two-new-coal-power-plants-per-week-in-2022.pdf.

45. Bloomberg News, "China's Energy Transition Will Be Slow and Steady," *Time,* October 17, 2022, https://time.com/6222566/china-fossil-fuel-plan-xi-jinping.

46. "Global Carbon Project: Briefing on Key Messages Global Carbon Budget 2023," Global Carbon Budget, December 4, 2023, 1–4, https://globalcarbonbudget.org/carbonbudget2023 (then navigate to "Key Messages").

47. "Global Carbon Project," 4; Pierre Friedlingstein, Michael O'Sullivan, Matthew W. Jones, Robbie M. Andrew, Dorothee C. E. Bakker, Judith Hauck, Peter Landschützer, et al., "Global Carbon Budget 2023," *Earth System Science Data* 15, no. 12 (December 5, 2023): 5301–5369, https://essd.copernicus.org/articles/15/5301/2023.

48. "Global Carbon Project," 6.

49. "Paris Agreement," United Nations, Department of Economic and Social Affairs, accessed January 14, 2024, https://sdgs.un.org/frameworks/parisagreement#background.

50. On natural gas leakage, see Hiroko Tabuchi, "Leaks Can Make Natural Gas as Bad for the Climate as Coal, a Study Says," *New York Times*, July 19, 2023, https://www.nytimes.com/2023/07/13/climate/natural-gas-leaks-coal-climate-change.html.

51. "Net Zero Roadmap: A Global Pathway to Keep the 1.5 °C Goal in Reach—2023 Update," International Energy Agency, September 2023, 16, https://iea.blob.core.windows.net/assets/9a698da4-4002-4e53-8ef3-631d8971bf84/NetZeroRoadmap_AGlobalPathwaytoKeepthe1.5CGoalinReach-2023Update.pdf.

52. "Heavy Dependence on Carbon Capture and Storage 'Highly Economically Damaging,' says Oxford Report," University of Oxford, December 4, 2023, https://www.smithschool.ox.ac.uk/news/heavy-dependence-carbon-capture-and-storage-highly-economically-damaging-says-oxford-report#:~:text=%E2%80%9CRelying%20on%20mass%20deployment%20of,at%20the%20Oxford%20Smith%20School.

53. "COP28 Agreement Signals 'Beginning of the End' of the Fossil Fuel Era," United Nations Climate Change, December 13, 2023, https://unfccc.int/news/cop28-agreement-signals-beginning-of-the-end-of-the-fossil-fuel-era.

54. Lawrence H. Summers, "A New Chance for the World Bank," *Project Syndicate*, October 10, 2022, https://www.project-syndicate.org/commentary/new-chance-expanded-mission-for-world-bank-by-lawrence-h-summers-2022-10.

55. Lawrence H. Summers and N. K. Singh, "The Multilateral Development Banks the World Needs," *Project Syndicate*, July 24, 2023, https://www.project-syndicate.org/commentary/world-bank-mdbs-how-to-triple-funding-and-make-more-effective-by-lawrence-h-summers-and-n-k-singh-2023-07. I was Summers's councillor at Treasury from 1999 to 2001.

56. "G7 Establishes Climate Club," Federal Ministry for Economic Affairs and Climate Action, December 12, 2022, https://www.bmwk.de/Redaktion/EN/Pressemitteilungen/2022/12/20221212-g7-establishes-climate-club.html.

57. Quoted in John Ruwitch, "China Accuses U.S. of Containment and Warns of Potential Conflict," *All Things Considered*, March 7, 2023, https://www.npr.org/2023/03/07/1161570798/china-accuses-u-s-of-containment-warns-of-potential-conflict.

58. Quoted in Michael Klare, "Climate Change Cooperation Cannot Be an 'Oasis' in U.S.-China Relations, Kerry Told," Committee for a SANE U.S.-China Policy, September 14, 2021, https://www.saneuschinapolicy.org/blog/climate-change-cooperation-cannot-be-an-oasis-in-us-china-relations-kerry-told.

59. Bonnie S. Glaser and Abigail Wulf, "China Global Podcast: China's Role in Critical Mineral Supply Chains," GMF—German Marshall Fund, August 2, 2023, https://email.gmfus.org/s/5baa40f7ee64e5ded79cea75cef33e1312f22ae2?frame=1.

60. Interview transcript and recording available at David Roberts, "Biden Sets Out to Supercharge Industrial Decarbonization," Volts, March 27, 2024, https://www.volts.wtf/p/biden-sets-out-to-supercharge-industrial.

61. *Net Zero Roadmap: A Global Pathway to Keep the 1.5°C Goal in Reach, 2023 Update* (Paris: International Energy Agency, September 2023), 13, https://iea.blob.core.windows.net/assets/9a698da4-4002-4e53-8ef3-631d8971bf84/NetZeroRoadmap_AGlobalPathwaytoKeepthe1.5CGoalinReach-2023Update.pdf.

62. Butler-Sloss and Bond, "The Eight Deadly Sins of Analyzing the Energy Transition."

63. Mekala Krishnan, Humayun Tai, Daniel Pacthod, Sven Smit, Tomas Nauclér, Blake Houghton, Jesse Noffsinger, and Dirk Simon, "The Path to Net Zero: A Guide to Getting It Right," McKinsey Sustainability, November 20, 2023, https://www.mckinsey.com/capabilities/sustainability/our-insights/an-affordable-reliable-competitive-path-to-net-zero#. See the link for full report.

64. Cary Funk and Alec Tyson, "Millennial and Gen Z Republicans Stand out from Their Elders on Climate and Energy Issues," Pew Research Center, June 24, 2020, https://www.pewresearch.org/short-reads/2020/06/24/millennial-and-gen-z-republicans-stand-out-from-their-elders-on-climate-and-energy-issues.

65. *World Energy Outlook 2023* (Paris: International Energy Agency, 2023), 91–92, 155–159, https://iea.blob.core.windows.net/assets/2b0ded44-6a47-495b-96d9-2fac0ac735a8/WorldEnergyOutlook2023.pdf; *World Energy Outlook 2021* (Paris: International Energy Agency, December 2021), 16, 25, 34–35, https://iea.blob.core.windows.net/assets/4ed140c1-c3f3-4fd9-acae-789a4e14a23c/WorldEnergyOutlook2021.pdf.

Index